T0155975

Graduate Texts in Mathematics   269

Graduate Texts in Mathematics

Graduate Texts in Mathematics bridge the gap between passive study and creative understanding, offering graduate-level introductions to advanced topics in mathematics. The volumes are carefully written as teaching aids and highlight characteristic features of the theory. Although these books are frequently used as textbooks in graduate courses, they are also suitable for individual study.

For further volumes:
http://www.springer.com/series/136

M. Scott Osborne

Locally Convex Spaces

 Springer

M. Scott Osborne
Department of Mathematics
University of Washington
Seattle, WA, USA

ISSN 0072-5285   ISSN 2197-5612 (electronic)
ISBN 978-3-319-34374-7   ISBN 978-3-319-02045-7 (eBook)
DOI 10.1007/978-3-319-02045-7
Springer Cham Heidelberg New York Dordrecht London

Mathematics Subject Classification: 46A03, 46-01, 46A04, 46A08, 46A30

Printed on acid-free paper

Springer is part of Springer Science+Business Media (www.springer.com)

Preface

Functional analysis is an exceptionally useful subject, which is why a certain amount of it is included in most beginning graduate courses on real analysis. For most practicing analysts, however, the restriction to Banach spaces is not enough. This book is intended to cover most of the general theory needed for application to other areas of analysis.

Most books on functional analysis come in one of two types: Either they restrict attention to Banach spaces or they cover the general theory in great detail. For the kind of courses I have taught, those of the first type don't cover a broad enough range of spaces, while those of the second type cover too much material for the time allowed. Don't misunderstand; there is plenty of interesting material in both kinds, and the interested party is invited to check them out. In fact, in this book, some side topics (e.g., topological groups or Mahowald's theorem) are treated, provided they don't go too far afield from the central topic. Also, some useful things are included that are hard to find elsewhere: Sect. 5.2 contains results that are not new but are not covered in most treatments (although Yosida [41] comes very close), for example.

The material here is based on a course I taught at the University of Washington. The course lasted one quarter and covered most of the material through Chap. 5. I was using Rudin [32] as a text. His early coverage of locally convex spaces is excellent, but he discusses dual spaces only for Banach spaces. I used his book to organize things in the course but had to expand considerably on the subject matter. (Rudin's book, by the way, is one of the few that does not fit the "two types" dichotomy above. Neither do Conway [7] or Reed and Simon [29].)

The prerequisite for this book is the Banach space theory typically taught in a beginning graduate real analysis course. The material in Folland [15], Royden [30], or Rudin [31] cover it all; Bruckner, Bruckner, and Thomson [6] cover everything except the Riesz representation theorem for general compact Hausdorff spaces, and that appears here only in an application in Sect. 5.5 and in Appendix C. There are some topological results that are needed, which may or may not have appeared in a beginning graduate real analysis course; these appear in Appendix A.

One more thing. Although this book is oriented toward applications, the beauty of the subject may appeal to you. If so, there is plenty out there to look at: Edwards [13], Grothendieck [16], Horvath [18], Kelley and Namioka [21], Schaefer [33], and Treves [36] are good reading on the general subject. Bachman and Narici [2], Narici and Beckenstein [27], and Schechter [34] do a nice job with Banach spaces. Wilansky [39] updates with more modern concepts the functional analysts have come up with (e.g., "webbed spaces"). Edwards [13], Phelps [28], Reed and Simon [29], and Treves [36] have plenty of material on the interaction of functional analysis with outside subjects. Also, Swartz [35] and Wong [40] are recommended. Finally, Dieudonne [10] gives a very readable history of the subject.

Finally, some "thank you's": To Prof. Garth Warner and Clayton Barnes, for comments and suggestions. To Owen Biesel, Nathaniel Blair-Stahn, Ryan Card, Michael Gaul, and Dustin Mayeda, who took the original course. And (big time!) to Mary Sheetz, who put the manuscript together. And a closing thank you to Elizabeth Davis, Richard Kozarek, Ronald Mason, Michael Mullins, Huong Pham, Vincent Picozzi, Betsy Ross, Chelsey Stevens, Megan Stewart, and L. William Traverso, without whose aid this manuscript would never have been completed.

Seattle, WA, USA   M. Scott Osborne

Contents

Chapter 1
Topological Groups

1.1 Point Set Topology

Every locally convex space is a topological group, that is, a group that is also a topological space in which the group operations (multiplication and inversion) are continuous. A large number of the most basic results about locally convex spaces are actually valid for any topological group and can be established in that context with only a little additional effort. Since topological groups are important in their own right, it seems worthwhile to establish these basic results in the context of topological groups.

While the reader is assumed to be familiar with basic point set topology, there are some twists that may or may not be familiar. These are not so important for topological groups (though they are handy), but they are crucial for dealing with locally convex spaces.

A notational point should be made before proceeding. If A is a subset of a topological space, then its closure will be denoted by A^-. This is because we will need complex numbers, and $\bar{a}$ will denote complex conjugation. Similarly, the interior of A will be denoted by $\text{int}(A)$. This is because "$A°$" has traditionally been assigned a special meaning in the context of locally convex spaces (it is called the "polar" of A).

There are basically three subjects to be discussed. The one most likely to already be familiar is the notion of a net. While locally convex spaces can be studied without this concept, some substitute (e.g., filters) would be necessary without them.

A net is basically a generalized sequence in which the natural numbers are replaced by a directed set.

Definition 1.1. A **directed set** is a pair $(D, \prec)$, where D is a nonempty set and $\prec$ is a binary relation on D subject to the following conditions:

(i) For all $\alpha \in D, \alpha \prec \alpha$.
(ii) For all $\alpha, \beta, \gamma \in D, \alpha \prec \beta$ and $\beta \prec \gamma$ implies $\alpha \prec \gamma$.
(iii) For all $\alpha, \beta \in D$, there exists $\gamma \in D$ such that $\alpha \prec \gamma$ and $\beta \prec \gamma$.

M.S. Osborne, *Locally Convex Spaces*, Graduate Texts in Mathematics 269,
DOI 10.1007/978-3-319-02045-7_1, © Springer International Publishing Switzerland 2014

Note that (i) and (ii) make D look like a partially ordered set; conspicuous by its absence is the antisymmetry condition. The lack of antisymmetry is important for a number of applications (see below) and does not affect things much. The crucial addition is condition (iii), which is what the word directed usually signifies.

For some reason, it has become traditional to denote the elements of a directed set with lowercase Greek letters.

Definition 1.2. A **net** in a topological space X is a function from D to X, where $(D, \prec)$ is a directed set. This function is usually denoted by $\alpha \mapsto x_\alpha$ (or $\alpha \mapsto y_\alpha$, or something similar). This net $\langle x_\alpha \rangle$ **converges** to x if the following happens: Whenever U is an open subset of X, with $x \in U$, then there exists $\alpha \in D$ such that

$$\forall \beta \in D : \alpha \prec \beta \Rightarrow x_\beta \in U.$$

Note: As usual, we sometimes refer to the directed set as D, rather than the more proper $(D, \prec)$. Similarly, "$\alpha \succ \beta$" means $\beta \prec \alpha$. Also, as above, the net is usually denoted by $\langle x_\alpha \rangle$ or $\langle x_\alpha : \alpha \in D \rangle$, and convergence to x is denoted by $x_\alpha \to x$, $\lim x_\alpha = x$, or $\lim_D x_\alpha = x$. Since the notion of a net is a generalization of the notion of a sequence (with $\mathbb{N}$ being replaced by D), this is consistent with standard terminology for sequences.

Example 1 (cf. Bear [3]). Given a bounded function $f : [a, b] \to \mathbb{R}$, the Riemann–Darboux integral can be defined as a net limit as follows. A *partition* P of $[a, b]$ is a finite sequence $a = x_0 < x_1 < \cdots < x_n = b$, with $P = \{x_0, x_1, \ldots, x_n\}$. That is, a partition is a finite set P, with $\{a, b\} \subset P \subset [a, b]$. A *tagging* T of the partition P above is a selection of points $\{t_1, \ldots, t_n\}$ for which $x_{j-1} \leq t_j \leq x_j$, and a *tagged partition* is an ordered pair (P, T) for which P is a partition and T is a tagging of P. The *Riemann Sum $S(P, T, f)$* for this tagged partition is the sum:

$$S(P, T, f) = \sum_{j=1}^{n} f(t_j)(x_j - x_{j-1}).$$

The directed set $(D, \prec)$ is the set of all tagged partitions of $[a, b]$, with $(P, T) \prec (P', T')$ when $P \subset P'$. Note: The "ordering" *ignores* the tagging and so is *not* antisymmetric. Darboux's version of the Riemann integral is defined as

$$\int_a^b f(x)dx = \lim_D S(P, T, f).$$

See Bear [3] for more details, including how to produce Lebesgue integrals as a net limit.

The following three facts are elementary and provide typical examples of how the flexibility in choosing D can be exploited.

Proposition 1.3. *Suppose X is a topological space. Then:*

(a) If $A \subset X$, then A^- is the set of limits of nets from A.
(b) X is Hausdorff if, and only if, convergent nets have unique limits.
(c) If Y is a topological space, and $f : X \to Y$ is a function, then f is continuous
    if, and only if, for any net $\langle x_\alpha \rangle$ in X: $\lim x_\alpha = x \Rightarrow \lim f(x_\alpha) = f(x)$.

Proof. (a1) Suppose $\langle x_\alpha \rangle$ is a net in A, and $x_\alpha \to x$. If U is any open neighborhood
of x, then there exists α s.t. $\beta \succ \alpha \Rightarrow x_\beta \in U$. In particular, $x_\alpha \in U \cap A$, so
$U \cap A \neq \emptyset$. This just says that x is adherent to A, so $x \in A^-$.
(a2) Suppose $x \in A^-$. Set

$$D = \{U : x \in U \text{ and } U \text{ is open}\}$$
$$U \succ V \Leftrightarrow U \subset V$$

D is directed. If $U \in D$, then $U \cap A \neq \emptyset$ since $x \in A^-$; let x_U be some
element chosen from $U \cap A$. (Yes, the axiom of choice is used here.) By
definition, $\langle x_U \rangle$ is a net in A, and $V \succ U \Rightarrow x_V \in V \subset U$, so $x_U \to x$.
(b1) Suppose X is Hausdorff. To show that nets have unique limits, suppose some
net $\langle x_\alpha \rangle$, defined on a directed set D, has (at least) two distinct limits x and y,
$x \neq y$. Let U and V be disjoint open neighborhoods of x and y, respectively.
Since $x_\alpha \to x$, there exists $\beta_1 \in D$ s.t. $\alpha \succ \beta_1 \Rightarrow x_\alpha \in U$. Since $x_\alpha \to y$,
there exists $\beta_2 \in D$ s.t. $\alpha \succ \beta_2 \Rightarrow x_\alpha \in V$. But D is directed, so there exists
$\gamma \in D$ s.t. $\gamma \succ \beta_1$ and $\gamma \succ \beta_2$, from which $x_\gamma \in U \cap V$. But $U \cap V = \emptyset$, a
contradiction.
(b2) Suppose X is not Hausdorff. We construct a net with (at least) two distinct
limits. Since X is not Hausdorff, there exist $x, y \in X$, $x \neq y$, such
that whenever U and V are open neighborhoods of x and y, respectively,
necessarily $U \cap V \neq \emptyset$. Set

$$D = \{(U, V) : x \in U, y \in V, U \text{ and } V \text{ both open}\};$$
$$(U, V) \prec (\tilde{U}, \tilde{V}) \Leftrightarrow (U \supset \tilde{U} \text{ and } V \supset \tilde{V}).$$

If $\alpha = (U, V)$, let x_α be some element chosen from $U \cap V$. Note that D *is*
directed: It is clearly partially ordered, and $(U, V) \prec (U \cap \tilde{U}, V \cap \tilde{V})$ and
$(\tilde{U}, \tilde{V}) \prec (U \cap \tilde{U}, V \cap \tilde{V})$. Furthermore, given an open neighborhood U of x,
$\gamma = (\tilde{U}, \tilde{V}) \succ (U, X) \Rightarrow x_\gamma \in \tilde{U} \subset U$, so by definition $x_\alpha \to x$. Similarly, if
V is an open neighborhood of y, then $\gamma = (\tilde{U}, \tilde{V}) \succ (X, V) \Rightarrow x_\gamma \in \tilde{V} \subset V$,
so $x_\alpha \to y$.
(c1) Suppose f is continuous, $\lim x_\alpha = x$, and U is open in Y with $f(x) \in U$.
Then $x \in f^{-1}(U)$, and $f^{-1}(U)$ is open, so there exists α such that $\beta \succ \alpha \Rightarrow$
$x_\beta \in f^{-1}(U) \Rightarrow f(x_\beta) \in U$. This is convergence.
(c2) Suppose f is not continuous. Then there exists $A \subset Y$, with A closed, such
that $f^{-1}(A)$ is not closed. By part (a), there is a convergent net $\langle x_\alpha \rangle$, with

$x_\alpha \in f^{-1}(A)$, for which $x = \lim x_\alpha \notin f^{-1}(A)$. But now $f(x) \notin A$, and $f(x_\alpha) \in A$, so part (a) says that "$\lim f(x_\alpha) = f(x)$" is impossible. □

Before leaving the subject of nets, there are three last "basic" considerations. Suppose $\langle x_\alpha \rangle$ is a net defined on a directed set D. A subset $D' \subset D$ is called **cofinal** if for all $\alpha \in D$, there exists $\beta \in D'$ such that $\beta \succ \alpha$. (*Note:* This usage seems to be firmly imbedded in the topological and algebraic literature, even though it is inconsistent with the technical meaning of the prefix "co" in the category theoretic literature.)

Proposition 1.4. *Suppose X is a topological space, $\langle x_\alpha : \alpha \in D \rangle$ is a net in X, and D' is cofinal in D. Then D' is directed, and $\lim_D x_\alpha = x \Rightarrow \lim_{D'} x_\alpha = x$.*

Proof. D' is directed: If $\alpha, \beta \in D'$, then $\alpha, \beta \in D$, so there exists $\gamma \in D$ with $\gamma \succ \alpha$ and $\gamma \succ \beta$. D' is cofinal, so there exists $\delta \in D'$ with $\delta \succ \gamma$, whence $\delta \succ \alpha$ and $\delta \succ \beta$.

$\lim_{D'} x_\alpha = x$ if $\lim_D x_\alpha = x$: If U is an open neighborhood of x, then there exists $\alpha \in D$ with $x_\beta \in U$ whenever $\beta \succ \alpha$. There exists $\gamma \in D'$ with $\gamma \succ \alpha$, since D' is cofinal. If $\delta \in D'$, and $\delta \succ \gamma$, then $\delta \succ \alpha$, so $x_\delta \in U$. □

Our next consideration is the notion of a cluster point of a net. This echoes the notion of a cluster point of a sequence: If $\langle x_\alpha : \alpha \in D \rangle$ is a net in a topological space X, and $x \in X$, then x is a **cluster point** of $\langle x_\alpha \rangle$ if the following happens: If U is an open neighborhood of x, and $\alpha \in D$, then there exists a $\beta \in D$ with $\beta \succ \alpha$ and $x_\beta \in U$. Note that a limit of a convergent net is a cluster point of that net (Exercise 1).

Proposition 1.5. *Suppose X is a compact topological space, and $\langle x_\alpha : \alpha \in D \rangle$ is a net in X. Then $\langle x_\alpha \rangle$ has a cluster point in X.*

Proof. For all $\alpha \in D$, set

$$A_\alpha = \{x_\beta : \beta \succ \alpha\} \text{ and}$$

$$C_\alpha = A_\alpha^-.$$

Observe that if $\alpha, \beta \in D$, then there exists $\gamma \in D$ for which $\gamma \succ \alpha$ and $\gamma \succ \beta$. Hence $A_\gamma \subset A_\alpha \cap A_\beta$, so that $C_\alpha \cap C_\beta$ is a closed set containing A_γ. Thus $C_\gamma \subset C_\alpha \cap C_\beta$. Since $x_\gamma \in A_\gamma \subset C_\gamma$, this shows that the family $\mathscr{F} = \{C_\alpha : \alpha \in D\}$ has the finite intersection property, since each member of $\mathscr{F}$ is nonempty, and the intersection of any two members (and hence any n members, by induction on n) of $\mathscr{F}$ contains a member of $\mathscr{F}$. Since X is compact, $\bigcap \mathscr{F} \neq \emptyset$, that is there exists $x \in X$ such that $x \in C_\alpha$ for every α. But that just means that whenever U is an open neighborhood of x, and $\alpha \in D$, necessarily $U \cap A_\alpha \neq \emptyset$ (since $x \in A_\alpha^-$). That is, there exists $x_\beta \in U$ for some $\beta \succ \alpha$. □

The final notion is that of a **subnet**. It is rather complicated, but it does arise in Sect. 5.7. Like a subsequence, a subnet involves a reparametrization of a net, but the manner in which this happens is much more general; so general, in fact, that a subnet of a sequence need not be a subsequence. Suppose $\langle x_\alpha : \alpha \in D \rangle$ is a net. A subnet is defined as follows. One has another directed set $(D', \leq)$ and a function $\alpha' \mapsto \varphi(\alpha')$ from D' to D for which (roughly speaking) $\varphi(\alpha') \to \infty$. That is, if $\alpha \in D$, then there exists a $\beta' \in D'$ for which $\gamma' \geq \beta'$ implies $\varphi(\gamma') \succ \alpha$. Note that the original net $\langle x_\alpha \rangle$ is not part of the definition yet; the condition only looks at D, D', and φ. The actual subnet is the function $\alpha' \mapsto x_{\varphi(\alpha')}$.

There are a few exercises at the end of this chapter that should illuminate this concept. The appearance of a subnet in Sect. 5.7 will be self-contained, but should be clearer after trying out these exercises.

The next major topic has come in the back door already. In the preceding, observe that the phrase "open neighborhood" was used consistently, rather than the simpler "neighborhood." That is because we do

NOT

assume that neighborhoods are open. This is *crucial* for functional analysis, although it does not seem to be so essential for topological groups per se. It is equally important NOT to require that a neighborhood base consist of open sets. [For those not familiar with the general notion, a set A is a **neighborhood** of p when $p \in \text{int}(A)$.] There are two kinds of bases for a topology $\mathscr{T}$ on X. A (*global*) *base* is a set $\mathscr{B} \subset \mathscr{T}$ such that

$$\forall\, x \in X, U \in \mathscr{T} \text{ with } x \in U : \exists\, B \in \mathscr{B} \text{ s.t. } x \in B \subset U.$$

Note that a global base always consists of open sets. Not so for neighborhood bases. If $x \in X$, then a *neighborhood base* $\mathscr{B}_x$ for x is a collection of subsets of X satisfying the following two conditions:

(i) Each $B \in \mathscr{B}_x$ is a neighborhood of x; that is, $x \in \text{int}(B)$.
(ii) If U is open, and $x \in U$, then there exists $B \in \mathscr{B}_x$ such that $B \subset U$.

Observe that if we are given a neighborhood base $\mathscr{B}_x$ at each point $x \in X$, then these $\mathscr{B}_x$'s do, in fact, determine the topology: If U is a subset of X with the property that for all $x \in U$ there exists $B \in \mathscr{B}_x$ such that $B \subset U$, then any such $x \in U$ satisfies $x \in \text{int}(B) \subset \text{int}(U)$, that is all points of U are interior points. Hence U is open. [This is why condition (i) is present.]

In functional analysis, there is a standard construction of neighborhood bases which is used again and again, and it automatically yields *closed* sets; in fact, sets that are closed in coarser topologies, a fact frequently exploited.

Observe that if $\mathscr{B}$ is a global base, then $\mathscr{B}_x = \{ B \in \mathscr{B} : x \in B \}$ is a local base at x. However, local bases in general will not amalgamate into a global base, since local bases do not need to consist of open sets.

There is a subtlety built into the preceding. Note that condition (ii) reads "$B \subset U$" and not "$\text{int}(B) \subset U$"; *all* of B has to fit inside U. This leads to our final topic of this section, the meaning of the adverb "locally."

Suppose X is a topological space, perhaps with other structure as well. Let [*adjective*] denote a property that "makes sense" for subsets of X; that is, given $A \subset X$, then either A is [*adjective*] or it is not. Purely topological examples abound: open, closed, compact, connected, disconnected, finite, etc. Normally, [*adjective*] will be an adjective, but sometimes it is a participle or a short phrase. *With one exception*, a space X has the property [*adjective*] **locally** (phrased as "X is locally [*adjective*]") when each point of X has a neighborhood base consisting of [*adjective*] sets.

This is consistent with the usual definition of "locally connected," even though that definition would really be stated here as "locally open-and-connected." See Exercise 7 at the end of this chapter.

The above definition of "locally" is not consistent with the traditional definition for "locally compact." A space is called *locally compact* when each point has a compact neighborhood, although some texts (e.g. Royden [30] and Rudin [31]) require that each point have a neighborhood with a closure that is compact. For Hausdorff spaces (the only ones we shall use the term "locally compact" for), they are equivalent, although in general they are not: Munkres [26, p. 185] goes so far as to say, "Our definition of local compactness has nothing to do with 'arbitrarily small' neighborhoods, so there is some question whether we should call it local compactness at all." However:

Proposition 1.6. *Suppose X is a Hausdorff space. Then the following are equivalent:*

(i) *Each point of X has a neighborhood base consisting of compact sets.*
(ii) *Each point of X has an open neighborhood with compact closure.*
(iii) *Each point of X has a compact neighborhood.*

Proof. Clearly (i) $\Rightarrow$ (iii) and (ii) $\Rightarrow$ (iii), whether X is Hausdorff or not, while (iii) $\Rightarrow$ (ii) for Hausdorff spaces by taking a compact (hence closed) neighborhood K of a point $x : x \in \text{int}(K)$, while $\text{int}(K)^- \subset K$, so that $\text{int}(K)^-$ is compact.

To show that (ii) or (iii) (either will do) imply (i), appeal to the standard fact that X is now T_3: If $x \in X$, and U is open, with $x \in U$, then there exist disjoint open sets V and W with $x \in V$ and $(X - U) \subset W$. If K is a compact neighborhood of x, then $K \cap V^-$ is a compact neighborhood of x (it contains $\text{int}(K) \cap V$), and $K \cap V^- \subset V^- \subset X - W \subset U$. $\square$

In general, (i) $\Rightarrow$ (iii) and (ii) $\Rightarrow$ (iii), and that is it; see the exercises at the end of this chapter.

For the record, we list the following standard definitions for a topological space X:

1. X is T_0 if and only if whenever $x, y \in X$, with $x \neq y$, there exists an open set U that contains one point only from $\{x, y\}$.
2. X is T_1 if and only if points are closed.
3. X is T_2 if and only if X is Hausdorff.

4. X is *regular* if and only if whenever $x \in X$ and $A \subset X$, with A closed and $x \notin A$, there exist disjoint open U and V with $x \in U$ and $A \subset V$.
5. X is T_3 if and only if X is regular and T_1.
6. X is *normal* if and only if whenever A and B are disjoint closed sets, there exist disjoint open sets U and V with $A \subset U$ and $B \subset V$.
7. X is T_4 if and only if X is normal and T_1.

We close this section with a result illustrating the utility of the adverbial approach to the meaning of "locally."

Proposition 1.7. *Suppose X is a topological space. Then X is regular if, and only if, X is locally closed.*

Proof. First, suppose X is regular, $x \in X$, and U is open, with $x \in U$. Then $A = X - U$ is closed and does not contain x, so there exists disjoint open V and W with $x \in V$ and $A \subset W$. But now $x \in V \subset V^- \subset X - W \subset X - A = U$, so V^- is a closed neighborhood of x which is contained in U.

Now suppose X is locally closed, and $x \in X$, $A \subset X$, with A closed and $x \notin A$. Then $X - A$ is open and $x \in X - A$, so there exists a closed neighborhood C of x with $C \subset X - A$. But now $x \in \text{int}(C)$ and $A \subset X - C$, while $\text{int}(C)$ and $X - C$ are disjoint open sets. □

1.2 Topological Groups: Neighborhood Bases

As noted earlier, a topological group G is a set endowed with two structures, a group structure and a topological structure. Specifically, G is both an abstract group and a topological space, with the two structures being compatible with each other. That is, the two maps

$$G \times G \to G : (x, y) \mapsto xy$$

$$G \to G : x \mapsto x^{-1}$$

are assumed to be continuous. Eventually, the Hausdorff condition will be imposed, but for now the above is all that will be assumed.

For further reading, two books (Dikranjan, Prodonov and Stayanov [11], and Husain [19]) are recommended.

Examples. (1) $(\mathbb{R}, +)$, usual topology on $\mathbb{R}$. (2) Any group, with the discrete topology. (3) Any group, with the indiscrete topology. (4) Matrix groups: The set of $n \times n$ invertible matrices with real (or complex) entries forms a topological group, where the topology is the usual Euclidean topology from $\mathbb{R}^{n^2}$ (or $\mathbb{C}^{n^2}$). (5) If X is a Banach space, then the set of all bounded linear bijections forms a group. (Inverses are bounded, thanks to the open mapping theorem.) In the operator norm topology, this produces a topological group. [Inversion is continuous because $T^{-1} - T_n^{-1} =$

$T^{-1}(T_n - T)T_n^{-1}$, so to show that $T_n \to T \Rightarrow T_n^{-1} \to T^{-1}$, it suffices to show that T_n^{-1} stays bounded. This happens because $\|T^{-1}\| \leq M \Rightarrow \|T(x)\| \geq M^{-1}\|x\|$, so once $\|T - T_n\| < M^{-1}/2$, one gets that $\|T_n(x)\| \geq (M^{-1}/2)\|x\|$, giving $\|T_n^{-1}\| \leq 2M$.]

NONEXAMPLES: (1) $\mathbb{R}$, with the "half-open interval topology," for which the set of all half-open intervals $[a, b)$ is a global base (this is sometimes called the "Sorgenfrey line"), is an example for which addition is continuous but negation is not, so $(\mathbb{R}, +)$ is not a topological group with this topology. (2) $\mathbb{Z}$, with the cofinite topology, is an example for which negation is continuous but addition is not, so $(\mathbb{Z}, +)$ is not a topological group with the cofinite topology.

Some things really are obvious. For example, multiplication is jointly continuous, hence is separately continuous. This just means that left multiplication and right multiplication are both continuous. These then automatically become homeomorphisms, since for example, $x \mapsto ax$ has $x \mapsto a^{-1}x$ as its inverse map. In particular, all inner automorphisms are also homeomorphisms. Inversion is also a homeomorphism. Also, a neighborhood base at the identity, e, can be either left multiplied or right multiplied by any $a \in G$ to get a neighborhood base at any point. Also, the opposite group, G^{op} with the same underlying set and topology but with multiplication reversed ($a * b = ba$) is also a topological group.

Suppose $\mathscr{B}_e$ is a neighborhood base at e. The following properties now must hold:

(i) If $B_1 \in \mathscr{B}_e$, then since inversion is continuous, there exists $B_2 \in \mathscr{B}_e$ such that $B_2^{-1} \subset B_1$. (Here, $B^{-1} = \{b^{-1} : b \in B\}$.)

(ii) If $B_1 \in \mathscr{B}_e$, then since multiplication is jointly continuous, there exist $B_2, B_3 \in \mathscr{B}_e$ for which $B_2 B_3 \subset B_1$. (Here $AB = \{ab : a \in A, b \in B\}$.)

(iii) If $B_1 \in \mathscr{B}_e$, then since inner automorphisms are homeomorphisms, for all $g \in G$ there exists $B_2 \in \mathscr{B}_e$ such that $gB_2g^{-1} \subset B_1$.

(iv) Finally, since the intersection of two open sets is open, for all $B_1, B_2 \in \mathscr{B}_e$ there exists $B_3 \in \mathscr{B}_e$ such that $B_3 \subset B_1 \cap B_2$.

Condition (iv) is what we would normally impose on any neighborhood base; the rest come from the topological group structure. It turns out that these conditions suffice to manufacture a topological group from an abstract group.

Proposition 1.8. *Suppose G is a group, and $\mathscr{B}_e$ is a nonempty collection of subsets of G; each containing the identity, e; satisfying:*

(i) For all $B_1 \in \mathscr{B}_e$ there exists $B_2 \in \mathscr{B}_e$ s.t. $B_2^{-1} \subset B_1$.

(ii) For all $B_1 \in \mathscr{B}_e$ there exist $B_2, B_3 \in \mathscr{B}_e$ s.t. $B_2 B_3 \subset B_1$.

(iii) For all $B_1 \in \mathscr{B}_e$ for all $g \in G$ there exists $B_2 \in \mathscr{B}_e$ s.t. $gB_2g^{-1} \subset B_1$.

(iv) For all $B_1, B_2 \in \mathscr{B}_e$ there exists $B_3 \in \mathscr{B}_e$ s.t. $B_3 \subset B_1 \cap B_2$.

Set

$$\mathscr{T} = \{U \subset G : \forall x \in U \, \exists B \in \mathscr{B}_e \text{ s.t. } xB \subset U\}.$$

Then $\mathcal{T}$ is a topology, and with this topology G is a topological group. Finally, for all $g \in G$, $g\mathcal{B}_e$ is a neighborhood base at g for this topology, as is $\mathcal{B}_e g$.

Proof. 1. $\mathcal{T}$ is a topology: Suppose $(U_\alpha : \alpha \in \mathcal{O}\}$ is a collection of members of $\mathcal{T}$. If x belongs to their union, U, then $x \in U_\alpha$ for some α, so $xB \subset U_\alpha \subset U$ for some $B \in \mathcal{B}_e$. If $U_1, U_2 \in \mathcal{T}$, and $x \in U_1 \cap U_2$, then there exist $B_1, B_2 \in \mathcal{B}_e$ with $xB_1 \subset U_1$ and $xB_2 \subset U_2$. there exists $B_3 \in \mathcal{B}_e$ with $B_3 \subset B_1 \cap B_2$, so $xB_3 \subset x(B_1 \cap B_2) = xB_1 \cap xB_2 \subset U_1 \cap U_2$. Finally, $\emptyset \in \mathcal{T}$ trivially, while $G \in \mathcal{T}$ since $\mathcal{B}_e$ is nonempty.

2. If $A \subset G$, then $\mathrm{int}(A) = \{x \in A:$ there exists $B \in \mathcal{B}_e$ s.t. $xB \subset A\}$: Set

$$U = \{x \in A : \exists B \in \mathcal{B}_e \text{ s.t. } xB \subset A\}.$$

Then $U \subset A$ by definition, while if V is open with $V \subset A$, necessarily for all $x \in V$ there exists $B \in \mathcal{B}_e$ with $xB \subset V \subset A$, so $V \subset U$. Hence if U is open, then it will be the largest open subset of A, and so will be $\mathrm{int}(A)$. While our list of properties (i)–(iv) do not appear to guarantee that U be open, the required property has snuck in via condition (ii).

U is open: Suppose $x \in A$, and $B_1 \in \mathcal{B}_e$ is such that $xB_1 \subset A$. By condition (ii), there exists $B_2, B_3 \in \mathcal{B}_e$ with $B_2 B_3 \subset B_1$. If $b \in B_2$, then $xbB_3 \subset xB_2 B_3 \subset xB_1 \subset A$, so $xb \in U$. In particular, $xB_2 \subset U$. So: Given $x \in U$, we manufacture B_2 with $xB_2 \subset U$, so U is open.

3. $\mathcal{B}_e$ is a neighborhood base at e: If $B \in \mathcal{B}_e$, then $eB = B \subset B$, so by the above, $e \in \mathrm{int}(B)$. If U is open, with $e \in U$, then by definition of $\mathcal{T}$ there exists $B \in \mathcal{B}_e$ with $B = eB \subset U$.

4. Multiplication is jointly continuous: Suppose U is open in G, and $xy \in U$. There exists $B_1 \in \mathcal{B}_e$ with $xyB_1 \subset U$. There exist $B_2, B_3 \in \mathcal{B}_e$ with $B_2 B_3 \subset B_1$. There exists $B_4 \in \mathcal{B}_e$ with $y^{-1}B_4 y \subset B_2$. Hence

$$xB_4 yB_3 = xyy^{-1}B_4 yB_3 \subset xyB_2 B_3 \subset xyB_1 \subset U.$$

In particular, $\mathrm{int}(xB_4) \cdot \mathrm{int}(yB_3) \subset U$. But $xB_4 \subset xB_4 \Rightarrow x \in \mathrm{int}(xB_4)$, while similarly $y \in \mathrm{int}(yB_3)$, so $\mathrm{int}(xB_4) \times \mathrm{int}(yB_3)$ is a neighborhood of (x, y) in $G \times G$ which multiplication maps into U.

5. Inversion is continuous: Suppose U is open, and $V = \{x \in G : x^{-1} \in U\}$. Suppose $x \in V$, so that $x^{-1} \in U$. There exists $B_1 \in \mathcal{B}_e$ such that $x^{-1}B_1 \subset U$. There exists $B_2 \in \mathcal{B}_e$ such that $xB_2 x^{-1} \subset B_1$. There exists $B_3 \in \mathcal{B}_e$ such that $B_3^{-1} \subset B_2$. Then

$$(xB_3)^{-1} = B_3^{-1}x^{-1} \subset B_2 x^{-1} = x^{-1}xB_2 x^{-1} \subset x^{-1}B_1 \subset U,$$

so $xB_3 \subset V$. Hence V is open. Finally,

6. For all $g \in G$, both $g \mathscr{B}_e$ and $\mathscr{B}_e g$ are neighborhood bases at g: Since G is now a topological group, both left and right multiplication by g are homeomorphisms, so this follows from (3). □

Examples. Proposition 1.8 yields a lot of examples. Note, for example, that if $\mathscr{B}_e$ consists of subgroups, then conditions (i) and (ii) are automatic, while (iii) and (iv) can be enforced by using a condition for which conjugate subgroups and intersections satisfy the condition. (1) $\mathscr{B}_e$ = all subgroups of finite index. This actually gives an interesting topological group structure on $(\mathbb{Z}, +)$. (2) By analogy with (1), let V be an infinite-dimensional vector space over a field, and set $\mathscr{B}_e$ = all subspaces of finite codimension. When the field is $\mathbb{R}$ or $\mathbb{C}$, this will yield a topological group structure that is *not* a topological vector space. (3) $S(\mathbb{N})$, the bijections of $\mathbb{N}$ with itself. $\mathscr{B}_e$ = the stabilizers of finite subsets of $\mathbb{N}$. This topology will provide a counterexample in Sect. 1.5.

Our next result concerns closures. It looks backwards, a fact that we can exploit.

Proposition 1.9. *Suppose G is a topological group, and $A \subset G$. Suppose $\mathscr{B}_e$ is a neighborhood base at the identity, e. Then*

$$A^- = \bigcap_{B \in \mathscr{B}_e} AB.$$

Proof. We show that if $x \in A^-$, then $x \in AB$ for all $B \in \mathscr{B}_e$; while if $x \notin A^-$, then $x \notin AB$ for some $B \in \mathscr{B}_e$.

First, suppose $x \in A^-$. If $B \in \mathscr{B}_e$, then $x(\operatorname{int}(B))^{-1} \cap A \neq \emptyset$, so there exists $a \in A$ and $b \in \operatorname{int}(B)$ such that $xb^{-1} = a$. But then $x = ab \in A\operatorname{int}(B) \subset AB$. Hence $x \in AB$ for all $B \in \mathscr{B}_e$.

Next, suppose $x \notin A^-$. Then $x \in G - A^-$, an open set, so $x^{-1} \in (G - A^-)^{-1}$, so $e \in (G - A^-)^{-1}x$. Hence there exists $B \in \mathscr{B}_e$ with $B \subset (G - A^-)^{-1}x$, so $Bx^{-1} \subset (G - A^-)^{-1}$, so $xB^{-1} \subset G - A^- \subset G - A$, that is $xB^{-1} \cap A = \emptyset$. But $x = ab \Rightarrow xb^{-1} = a$, so $x \in AB \Rightarrow xB^{-1} \cap A \neq \emptyset$. Hence $x \notin AB$ for this B. □

Corollary 1.10. *If G is a topological group, and $\mathscr{B}_e$ is a neighborhood base at the identity e, then $\{B^- : B \in \mathscr{B}_e\}$ is also a neighborhood base at e. In particular, G is regular.*

Proof. If $B \in \mathscr{B}_e$, then $e \in \operatorname{int}(B) \subset \operatorname{int}(B^-)$ since $B \subset B^-$. It remains to show that any open set U, with $e \in U$, there is a $B \in \mathscr{B}_e$ with $B^- \subset U$. But there exists $B_1 \in \mathscr{B}_e$ with $B_1 \subset U$, since $\mathscr{B}_e$ is a neighborhood base; while there exist $B_2, B_3 \in \mathscr{B}_e$ with $B_2 B_3 \subset B_1$, by condition (iii) on neighborhood bases at e. But $B_2^- \subset B_2 B_3$ by Proposition 1.9, so

$$B_2^- \subset B_2 B_3 \subset B_1 \subset U.$$

This shows that $\mathscr{B}_e^- = \{B^- : B \in \mathscr{B}_e\}$ is also a neighborhood base at e, so for all $g \in G$, $g\mathscr{B}_e^-$ is a neighborhood base at g consisting of closed sets. Since G is locally closed, it is regular (Proposition 1.7). $\qquad\square$

Corollary 1.11. *Suppose G is a topological group, with identity e, and $\mathscr{B}_e$ is a neighborhood base at e. Then the following are equivalent*

(i) G *is* T_0.
(ii) G *is* T_1.
(iii) G *is* T_2.
(iv) G *is* T_3.
(v) $\{e\}$ *is closed.*
(vi) $\bigcap\limits_{B\in\mathscr{B}_e} B = \{e\}$.

Proof. We know that

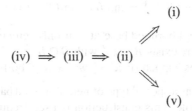

for general topological reasons. Corollary 1.10 says that (ii) $\Rightarrow$ (iv), since regular $+T_1 = T_3$. Also, (v) $\Rightarrow$ (vi) $\Rightarrow$ (ii), since by Proposition 1.9:

$$\text{Given (v): } \{e\} = \{e\}^- = \bigcap_{B\in\mathscr{B}_e} eB = \bigcap_{B\in\mathscr{B}_e} B; \text{ and}$$

$$\text{Given (vi): } \{g\}^- = \bigcap_{B\in\mathscr{B}_e} gB = g(\bigcap_{B\in\mathscr{B}_e} B) = g\{e\} = \{g\}.$$

The proof is completed by showing that (i) $\Rightarrow$ (v). This is done by showing that if $g \in G$ and $g \neq e$, then $g \notin \{e\}^-$. Since we are only assuming T_0, there are two cases:

1. There exists open U with $g \in U$ and $e \notin U$. Then $\{e\} \subset G - U$, a closed set, so $\{e\}^- \subset G - U$. But $g \in U$, so $g \notin \{e\}^-$.
2. There exists open U with $e \in U$ and $g \notin U$. Then there exists $B \in \mathscr{B}_e$ with $B \subset U$, so

$$g \notin U \supset B \supset \bigcap_{B'\in\mathscr{B}_e} eB' = \{e\}^-.$$

$\qquad\square$

So far, our neighborhood bases at e have been unrestricted. It is often helpful to have bases that are more restricted. The easy, general case, is as follows:

Proposition 1.12. *Suppose G is a topological group, with identity e. Then there is a neighborhood base $\mathscr{B}_e$ at e such that for all $B \in \mathscr{B}_e : B = B^{-1}$. Furthermore, the members of $\mathscr{B}_e$ may also be assumed to all be open, or they may be assumed to all be closed.*

Proof. Start with any neighborhood base $\mathscr{B}'_e$ at e; even $\{U : U \text{ open}, e \in U\}$ will do. Set

$$\mathscr{B}_e = \{B \cap B^{-1} : B \in \mathscr{B}'_e\}.$$

Since inversion is a homeomorphism, $\mathscr{B}_e$ will consist of open sets if $\mathscr{B}'_e$ did, and $\mathscr{B}_e$ will consist of closed sets if $\mathscr{B}'_e$ did. Also, the members $B_\bullet$ of $\mathscr{B}_e$ satisfy, $B_\bullet = B_\bullet^{-1}$, and for all $B \in \mathscr{B}'_e$,

$$e \in \text{int}(B) \cap (\text{int}(B))^{-1} = \text{int}(B) \cap \text{int}(B^{-1}) = \text{int}(B \cap B^{-1}).$$

To show that $\mathscr{B}_e$ is a neighborhood base at e, it only remains to show that if U is open and $e \in U$, then there exists $B \in \mathscr{B}'_e$ with $B \cap B^{-1} \subset U$. But this is automatic when $B \subset U$, which does happen since $\mathscr{B}'_e$ was a neighborhood base. $\square$

Our final topic concerns a special type of neighborhood base at e that exists when the group is first countable. This construction is used frequently to get at various results.

Theorem 1.13. *Suppose G is a first countable topological group, and U is an open set containing the identity, e. Then there exists a neighborhood base $\mathscr{B}_e = \{B_1, B_2, B_3, \ldots\}$ at e such that:*

(a) $B_1 \subset U$, and
(b) For all $j : B_j = B_j^{-1} \supset B_{j+1}^2$.

Furthermore, all the sets B_j may be assumed to be open, or all the sets B_j may be assumed to be closed.

Proof. Start with a countable neighborhood base $\mathscr{B}'_e$ at e, and manufacture $\mathscr{B}''_e = \{B \cap B^{-1} : B \in \mathscr{B}_e\}$; $\mathscr{B}''_e$ is also countable. It is a neighborhood base by the proof of Proposition 1.12. If you want open sets, set $\mathscr{B}'''_e = \{\text{int}(B) : B \in \mathscr{B}''_e\}$; if you want closed sets, set $\mathscr{B}'''_e = \{B^- : B \in \mathscr{B}''_e\}$. The latter is a neighborhood base via Corollary 1.10. $\mathscr{B}_e$ will be a subset of $\mathscr{B}'''_e$, so it will be countable, and will consist of open or closed sets if $\mathscr{B}'''_e$ did.

Choose any $B_1 \in \mathscr{B}'''_e$ with $B_1 \subset U$. Enumerate $\mathscr{B}'''_e$ as $B_1 = B(1)$, $B(2), \ldots$, and define B_j recursively as follows:

Given B_j, choose $B^{(1)} \in \mathscr{B}'''_e$ with $B^{(1)} \subset B_j \cap B(j+1)$. Choose $B^{(2)}$, $B^{(3)} \in \mathscr{B}'''_e$ with $B^{(2)} B^{(3)} \subset B^{(1)}$.
Choose $B_{j+1} \in \mathscr{B}'''_e$ with $B_{j+1} \subset B^{(2)} \cap B^{(3)}$.

Observe that $B_{j+1}^2 \subset B^{(2)} B^{(3)} \subset B^{(1)} \subset B_j$. Also, if V is open, with $e \in V$, then there exists $B(k) \in \mathscr{B}_e'''$ with $B(k) \subset V$, so that:

1. If $k = 1$, then $B_1 = B(1) \subset V$.
2. If $k = j + 1$, then

$$B_{j+1} \subset B^{(2)} \subset B^{(1)} \subset B(j+1) \subset V.$$

This just shows that $\mathscr{B}_e = \{B_1, B_2, \ldots\}$ is also a neighborhood base at e. $\qquad \square$

1.3 Set Products

We already have one result concerning products, namely, the formula for closures in Proposition 1.9. We will need a lot more. Some simple observations first.

1. A and B are subsets of any group, set $A^{-1} = \{a^{-1} : a \in A\}$ and $AB = \{ab : a \in A, b \in B\}$. The following rules are direct:

   (i) $AB = \bigcup_{a \in A} aB = \bigcup_{b \in B} Ab$
   (ii) $(AB)^{-1} = B^{-1}A^{-1}$
   (iii) $A(BC) = (AB)C = \{abc : a \in A, b \in B, c \in C\}$.
   (iv) $A(B \cap C) \subset AB \cap AC; (A \cap B)C \subset AC \cap BC$.
   (v) $A(B \cup C) = AB \cup AC; (A \cup B)C = AC \cup BC$.
   (vi) $x \in AB \Leftrightarrow xB^{-1} \cap A \neq \emptyset \Leftrightarrow A^{-1}x \cap B \neq \emptyset$.

Some of these look like conjuring tricks. In fact, the first time you see rules like these you *should* be suspicious. Rule (ii), for example, comes from properties of quantifiers:

$$x \in A(BC) \Leftrightarrow \exists a \in A, y \in BC \text{ with } x = ay$$

$$\Leftrightarrow \exists a \in A, \exists b \in B, \exists c \in C \text{ with } x = abc$$

$$\Leftrightarrow \exists c \in C, \exists a \in A, \exists b \in B \text{ with } x = abc$$

$$\Leftrightarrow \exists c \in C, \exists z \in AB \text{ with } x = zc$$

$$\Leftrightarrow x \in (AB)C.$$

As for (vi), note that $x = ab \Leftrightarrow xb^{-1} = a \Leftrightarrow a^{-1}x = b$, no matter *what* x, a, and b are. Hit this with some existential quantifiers, and you get (vi). The others work out just as quickly and are left as an exercise.

2. The topological properties of products seem to either come easily or with difficulty, with no in-between. The early ones are:

(i) If A or B is open, then AB is open, since rule (i) writes AB as a union of open sets.

(ii) If A and B are compact, then AB, as the continuous image of $A \times B$ under multiplication, is compact.

To get much further, we need some kind of uniform separation result. The following result covers this.

Theorem 1.14. *Suppose G is a topological group, and suppose $\mathscr{B}_e$ is a neighborhood base at the identity, e. Suppose A is a closed subset of G and K is a compact subset of G, with $A \cap K = \emptyset$. Then there exists $B \in \mathscr{B}_e$ such that $AB \cap KB = \emptyset$.*

Proof. The proof comes in two steps.

Step 1. There exists $B \in \mathscr{B}_e$ for which $(AB)^- \cap K = \emptyset$. Suppose not. Set $C_B = (AB)^- \cap K$ and $\mathscr{C} = \{C_B : B \in \mathscr{B}_e\}$. $\mathscr{C}$ is a collection of (relatively) closed, nonempty subsets of K. If $B_1, B_2 \in \mathscr{B}_e$, then there exists $B_3 \in \mathscr{B}_e$ with $B_3 \subset B_1 \cap B_2$, so

$$AB_3 \subset A(B_1 \cap B_2) \subset AB_1 \cap AB_2 \subset (AB_1)^- \cap (AB_2)^-,$$

so that $(AB_3)^-$, as the smallest closed set containing AB_3, satisfies $(AB_3)^- \subset (AB_1)^- \cap (AB_2)^-$. Intersecting with K, we get that $C_{B_3} \subset C_{B_1} \cap C_{B_2}$. An easy induction on n now shows that the intersection of n members of $\mathscr{C}$ contains a member of $\mathscr{C}$, and so is nonempty. That is, $\mathscr{C}$ has the finite intersection property. Since K is compact, this means that there exists $x \in K$ with $x \in C_B$ for all $B \in \mathscr{B}_e$.

This cannot happen. Since $x \in K$, and A is closed, by Proposition 1.9:

$$x \notin A = A^- = \bigcap_{B \in \mathscr{B}_e} AB.$$

That is, there exists $B_1 \in \mathscr{B}_e$ with $x \notin AB_1$. There exist $B_2, B_3 \in \mathscr{B}_e$ with $B_2 B_3 \subset B_1$, so that (again by Proposition 1.9):

$$x \notin AB_1 \supset A(B_2 B_3) = (AB_2)B_3 \supset (AB_2)^-, \text{ a contradiction.}$$

Step 2. The theorem is true. Choose $B_1 \in \mathscr{B}_e$ for which $(AB_1)^- \cap K = \emptyset$. Choose $B_2, B_3 \in \mathscr{B}_e$ for which $B_2 B_3 \subset B_1$. Choose $B_4 \in \mathscr{B}_e$ for which $B_4 \subset B_3^{-1}$. Choose $B \in \mathscr{B}_e$ for which $B \subset B_2 \cap B_4$. That's our B.

Suppose $x \in AB \cap KB$, that is $x = ab = g\tilde{b}$, with $a \in A$, $g \in K$, and $b, \tilde{b} \in B$. Then $g = ab\tilde{b}^{-1}$. But $\tilde{b} \in B \subset B_4 \subset B_3^{-1}$, so $\tilde{b}^{-1} \in B_3$, while $b \in B \subset B_2$, so $g = ab\tilde{b}^{-1} \in AB_2 B_3 \subset AB_1$, which is disjoint from K. This is a contradiction. $\square$

Corollary 1.15. *Suppose G is a topological group, and suppose A is a closed subset, and K is a compact subset. Then AK and KA are closed subsets of G.*

Proof. AK is closed: It suffices to show that $x \notin AK \Rightarrow x \notin (AK)^-$. Suppose $x \notin AK$. Then $A^{-1}x \cap K = \emptyset$ by property (vi) of set products. A is closed, so A^{-1} is closed, as is $A^{-1}x$. By Theorem 1.14, there exists $B \in \mathcal{B}_e$ for which $A^{-1}xB \cap KB = \emptyset$; in particular, $A^{-1}xB \cap K = \emptyset$. If some $xb = ag$ ($b \in B, a \in A, g \in K$), then $a^{-1}xb = g \in A^{-1}xB \cap K = \emptyset$, a contradiction, so $xB \cap AK = \emptyset$. Since xB is a neighborhood of x which is disjoint from AK, $x \notin (AK)^-$.

As for KA, use the above:

$$KA = (A^{-1}K^{-1})^{-1}, \text{ and } A^{-1}K^{-1} \text{ is closed.}$$

$\square$

These results apply primarily to quotients, the subject of the next section, but one loose end remains.

Proposition 1.16. *Suppose A and B are subsets of a topological group G. Then $(A^-)(B^-) \subset (AB)^-$.*

Proof. If $a \in A$, set $L_a(x) = ax$. L_a is continuous, so $L_a^{-1}((AB)^-)$ is closed. $L_a(B) = aB \subset AB \subset (AB)^-$, so $B \subset L_a^{-1}((AB)^-)$: thus, $B^- \subset L_a^{-1}((AB)^-)$. Hence $a(B^-) \subset (AB)^-$. Letting a float, $A(B^-) \subset (AB)^-$.

If $b \in B^-$, set $R_b(x) = xb$. R_b is continuous, so $R_b^{-1}((AB)^-)$ is closed. By the above, $Ab \subset (AB)^-$, so $R_b(A) \subset (AB)^-$, that is $A \subset R_b^{-1}((AB)^-)$; thus $A^- \subset R_b^{-1}((AB)^-)$. Hence $(A^-)b \subset (AB)^-$. Letting b float, $(A^-)(B^-) \subset (AB)^-$.

$\square$

1.4 Constructions

There are various ways of making new topological groups from old ones. Here, we shall stick to those which will be relevant to locally convex spaces. Products come first.

The product topology is usually characterized as the coarsest topology on the set-theoretic product for which the projections onto the factors are continuous. There is an alternate characterization, which comes from category theory, which is more useful here. The underlying theme will recur two more times in this section, and another time in the exercises. It is not particularly difficult, but it is peculiar, and takes getting used to.

We need a lemma first.

Lemma 1.17. *Suppose X and Y are topological spaces, $f : X \to Y$ is a function, and $\mathcal{B}$ is a subbase for the topology of Y. If $f^{-1}(B)$ is open in X for all $B \in \mathcal{B}$, then f is continuous.*

Proof. Suppose $f^{-1}(B)$ is open in X for all $B \in \mathcal{B}$. Let $\mathcal{B}_0$ denote the set of all finite intersections of members of $\mathcal{B}$. Since $f^{-1}(B_1 \cap \cdots \cap B_n) = f^{-1}(B_1) \cap \cdots \cap f^{-1}(B_n)$: $f^{-1}(B)$ is open in X for all $B \in \mathcal{B}_0$. But $\mathcal{B}_0$ is a base for the topology of Y, so any open subset of Y is a union of members of $\mathcal{B}_0$. If $U = \cup B_i$, then $f^{-1}(U) = f^{-1}(\cup B_i) = \cup f^{-1}(B_i)$ is open in X.

$\square$

Now. to the point:

Theorem 1.18. *Suppose* $\langle X_i : i \in \mathcal{I} \rangle$ *is a family of topological spaces. Then the product topology on* $\prod X_i$ *is the unique topology with the following property: Whenever Y is a topological space, and*

$$\mathbf{f} = \langle f_i \rangle : Y \to \prod_{i \in \mathcal{I}} X_i$$

is a function, then $\mathbf{f}$ *is continuous if, and only if, every* $f_i : Y \to X_i$ *is continuous.*

Proof. Let $\pi_{i_0} : \prod X_i \to X_{i_0}$ be a projection. If $\mathbf{f}$ is continuous, then each $f_{i_0} = \pi_{i_0} \circ \mathbf{f}$ is continuous. On the other hand, if each f_i is continuous, then since

$$\mathbf{f}^{-1} \left(\prod_{i \in \mathcal{I}} \left\{ \begin{matrix} X_i \text{ if } i \neq i_0 \\ U_{i_0} \text{ if } i = i_0 \end{matrix} \right\} \right) = f_{i_0}^{-1}(U_{i_0}),$$

$\mathbf{f}^{-1}(U)$ when U is any subbase element, is open, so $\mathbf{f}$ is continuous by Lemma 1.17. Thus, the product topology has the property in question.

Now for uniqueness. Here's where things get strange. Suppose $\mathcal{T}$ is any topology on the set $\prod X_i$ with the property that functions from some topological space Y to $\prod X_i$ are continuous exactly when their coordinate functions are continuous. Then the identity map from $(\prod X_i, \mathcal{T})$ to itself is certainly continuous, so its coordinate functions, the projections π_i, must also be continuous.

Now suppose $\mathcal{T}_1$ and $\mathcal{T}_2$ are any two topologies on $\prod X_i$ with the property that maps to $\prod X_i$ are continuous exactly when their coordinate functions are continuous. Consider the identity map from $(\prod X_i, \mathcal{T}_1)$ to $(\prod X_i, \mathcal{T}_2)$. The coordinate functions are the projections to the factors, and these are now known to be continuous, so this identity map is continuous. Reversing the roles of $\mathcal{T}_1$ and $\mathcal{T}_2$, this identity map must be a homeomorphism, so $\mathcal{T}_1 = \mathcal{T}_2$. $\square$

Corollary 1.19. *Suppose* $\langle X_{ij}, (i, j) \in \mathcal{I} \times \mathcal{J} \rangle$ *is a family of topological spaces parametrized by a set product* $\mathcal{I} \times \mathcal{J}$. *Then the set bijections*

$$\prod_{i \in \mathcal{I}} \prod_{j \in \mathcal{J}} X_{i,j} \approx \prod_{(i,j) \in \mathcal{I} \times \mathcal{J}} X_{i,j} \approx \prod_{j \in \mathcal{J}} \prod_{i \in \mathcal{I}} X_{i,j}$$

are homeomorphisms.

Proof. The underlying idea is pretty simple: Show that the product-of-product topology on $\prod \prod X_{i,j}$ has the property specified in Theorem 1.18. Since the situation is symmetric, we use the first bijection. Suppose Y is a topological space, and $\mathbf{f} = \langle f_{i,j} \rangle$ is a function from Y to $\prod_{\mathcal{I}} \prod_{\mathcal{J}} X_{i,j}$. Let $\mathbf{f}_i$ denote the ith partial coordinate function,

$$\mathbf{f}_i : Y \to \prod_{j \in \mathcal{J}} X_{i,j}.$$

That is, $\mathbf{f}_i = \langle f_{i,j} : j \in \mathscr{J} \rangle$, with i fixed. Then $\mathbf{f}$ is continuous $\Leftrightarrow$ for all $i \in \mathscr{I}$, $\mathbf{f}_i$ is continuous $\Leftrightarrow$ for all $i \in \mathscr{I}$ for all $j \in \mathscr{J}$ $f_{i,j}$ is continuous. Hence $\mathbf{f}$ is continuous $\Leftrightarrow$ all $f_{i,j}$ are continuous, $i \in \mathscr{I}$ and $j \in \mathscr{J}$. $\qquad\square$

Corollary 1.20. *Suppose $\langle G_i, i \in \mathscr{I} \rangle$ is a family of topological groups. Then with the product topology, $\prod G_i$ is a topological group.*

Proof. 1. *Inversion is continuous:* Since $\langle x_i \rangle^{-1} = \langle x_i^{-1} \rangle$, the i_0th coordinate function of $g \mapsto g^{-1}$ is $\langle x_i \rangle \mapsto x_{i_0}^{-1}$, that is it is the composite $g \mapsto \pi_{i_0}(g) \mapsto \pi_{i_0}(g)^{-1}$, a composite of two continuous maps.

2. *Multiplication is continuous:* Let $\mu_i : G_i \times G_i \to G_i$ denote the multiplication map. Then since $\langle x_i \rangle \cdot \langle y_i \rangle = \langle x_i y_i \rangle$, the i_0th coordinate of the multiplication map is

$$\left(\prod_{i \in \mathscr{I}} G_i\right) \times \left(\prod_{i \in \mathscr{I}} G_i\right) \approx \prod_{i \in \mathscr{I}}(G_i \times G_i) \xrightarrow{\text{proj.}} G_{i_0} \times G_{i_0} \xrightarrow{\mu_{i_0}} G_{i_0}$$
$$\uparrow$$
$$\text{Cor. 1.19}$$

which is continuous. $\qquad\square$

Note that the underlying idea of both parts is the same: The operation, with values in $\prod G_i$, is continuous because its coordinate functions are continuous. The diagram that illustrates this for inversion is

$$\prod_{i \in \mathscr{I}} G_i \xrightarrow{\text{proj.}} G_{i_0} \xrightarrow{\text{inv.}} G_{i_0}.$$

Now to subgroups. Again, it all starts with a topology result.

Proposition 1.21. *Suppose $(X, \mathscr{T}_X)$ is a topological space, and A is a subspace, with subspace topology $\mathscr{T}_A$. Then $\mathscr{T} = \mathscr{T}_A$ is the unique topology on A with the following property: Whenever Y is a topological space, and $f : Y \to A$ is a function, then $f : Y \to (X, \mathscr{T}_X)$ is continuous if, and only if, $f : Y \to (A, \mathscr{T})$ is continuous.*

Proof. If $f : Y \to A$ is a function, then for all $U \subset X : f^{-1}(U) = f^{-1}(U \cap A)$. But $f^{-1}(U)$ is a typical inverse image of a member of $\mathscr{T}_X$, while $f^{-1}(U \cap A)$ is a typical inverse image of a member of $\mathscr{T}_A$, so $\mathscr{T} = \mathscr{T}_A$ does have the property in question.

Now for uniqueness, which follows the routine from the proof of Theorem 1.18. Suppose $\mathscr{T}$ is a topology on A with the property in question. Let $Y = (A, \mathscr{T})$, and $f = $ identity map. This f is continuous from Y to $(A, \mathscr{T})$, so it must be continuous from Y to $(X, \mathscr{T}_X)$. That is, the inclusion $(A, \mathscr{T}) \hookrightarrow (X, \mathscr{T}_X)$ will automatically be continuous.

Now suppose $\mathcal{T}_1$ and $\mathcal{T}_2$ are two topologies on A with the property in question. Consider the identity map: $(A, \mathcal{T}_1) \to (A, \mathcal{T}_2)$. The map from $(A, \mathcal{T}_1) \to (X, \mathcal{T}_X)$ was just verified to be continuous, so $(A, \mathcal{T}_1) \to (A, \mathcal{T}_2)$ is continuous. Reversing the roles of $\mathcal{T}_1$ and $\mathcal{T}_2$, this identity map must be a homeomorphism, so $\mathcal{T}_1 = \mathcal{T}_2$.

$\square$

Corollary 1.22. *Suppose G is a topological group, and H is a subgroup. Then with the induced topology, H is a topological group.*

Proof.

$$H \times H \hookrightarrow G \times G \xrightarrow{\quad \text{multiplication} \quad} G$$

and

$$H \hookrightarrow G \xrightarrow{\quad \text{inversion} \quad} G$$

are continuous and take values in H. Apply Proposition 1.21. $\square$

Quotients take a bit more discussion. The topological aspects do not require the subgroup to be normal.

Suppose G is a topological group, and H is a subgroup. Let G/H denote the set of left cosets of H. Set

$$\mathcal{T}_{G/H} = \{UH/H : U \text{ open in } G\}.$$

Theorem 1.23. *Suppose G is a topological group, and H is a subgroup. Then:*

(a) *$\mathcal{T}_{G/H}$ is a topology on G/H.*
(b) *The natural projection $\pi : G \to G/H$ is both continuous and open.*
(c) *If Y is a topological space, and $f : G/H \to Y$ is a function, then f is continuous if, and only if, $f \circ \pi : G \to Y$ is continuous.*
(d) *$\mathcal{T}_{G/H}$ is the only topology on G/H with the property described in (c): Suppose $\mathcal{T}$ is a topology on G/H with the property that whenever Y is a topological space and $f : G/H \to Y$ is a function, then $f : (G/H, \mathcal{T}) \to Y$ is continuous $\Leftrightarrow f \circ \pi : G \to Y$ is continuous. Then $\mathcal{T} = \mathcal{T}_{G/H}$.*
(e) *If Y is a topological space, and $f : G/H \to Y$ is a function, then f is an open map if, and only if, $f \circ \pi : G \to Y$ is an open map.*
(f) *If $\mathcal{B}_e$ is a neighborhood base at the identity $e \in G$, then $\{BgH/H : B \in \mathcal{B}_e\}$ is a neighborhood base at gH for $\mathcal{T}_{G/H}$.*
(g) *If H is closed in G, then G/H is Hausdorff.*
(h) *If H is a normal subgroup of G, then with the topology $\mathcal{T}_{G/H}$, G/H is a topological group.*
(i) *If $\langle G_i, i \in \mathcal{I} \rangle$ is a family of topological groups, and H_i is a subgroup of G_i for each i, then the natural bijection*

$$\left(\prod_{i\in\mathcal{I}} G_i\right)\Big/\left(\prod_{i\in\mathcal{I}} H_i\right) \approx \prod_{i\in\mathcal{I}}(G_i/H_i)$$

is a homeomorphism.

Proof. These nine properties are not proved in the order given, but (a) definitely comes first. Let $\pi : G \to G/H$ be the natural map.

(a) $\emptyset = \emptyset H/H \in \mathcal{T}_{G/H}$ and $G/H = GH/H \in \mathcal{T}_{G/H}$. $\cup U_i H/H = \cup\pi(U_i) = \pi(\cup U_i) = (\cup U_i)H/H$ since direct images preserve unions. Direct images do not preserve intersections, though.

Suppose U and V are open in G. Then so are UH and VH. Clearly, $(UH \cap VH)H/H \subset UH/H \cap VH/H$. Suppose $xH \in UH/H \cap VH/H$. Then $xH \subset UH$, so $x \in UH$; similarly, $x \in VH$. Hence $x \in UH \cap VH$, so $xH \in (UH \cap VH)H/H$. Thus $UH/H \cap VH/H = (UH \cap VH)H/H \in \mathcal{T}_{G/H}$.

(b) Since $\pi(U) = UH/H$, π maps the topology of G onto $\mathcal{T}_{G/H}$. Since it takes values in $\mathcal{T}_{G/H}$, π is open. But $\pi^{-1}(UH/H) = UH$ is also open, so π is continuous.

(c) Suppose $f : G/H \to Y$ is a function, where Y is a topological space. If f is continuous, then $f \circ \pi$ is continuous since π is continuous. If $f \circ \pi$ is continuous, and U is open in Y, then $(f \circ \pi)^{-1}(U)$ is open in G, so $\pi\left((f \circ \pi)^{-1}(U)\right)$ is open in G/H since π is an open map. But $f^{-1}(U) = \pi\left((f \circ \pi)^{-1}(U)\right)$:

$$xH \in f^{-1}(U) \Leftrightarrow f(xH) \in U \Leftrightarrow f \circ \pi(x) \in U$$

$$\Leftrightarrow x \in (f \circ \pi)^{-1}(U) \Leftrightarrow xH \subset (f \circ \pi)^{-1}(U).$$

The last comes from the fact that π is constant on left cosets of H, so that $(f \circ \pi)^{-1}(U) = \pi^{-1}(f^{-1}(U))$ is a union of left cosets of H. But with this in mind,

$$xH \subset (f \circ \pi)^{-1}(U) \Leftrightarrow xH \in \pi\left((f \circ \pi)^{-1}(U)\right).$$

(d) Suppose $\mathcal{T}$ is a topology on G/H with the property specified. Set $Y = G/H$ with this topology, and $f = $ identity. Then f is continuous, so $f \circ \pi = \pi$ must be continuous.

Suppose $\mathcal{T}_1$ and $\mathcal{T}_2$ are two topologies on G/H with the property specified. Use $\mathcal{T}_1$ on G/H, and let Y be G/H with topology $\mathcal{T}_2$, $f = $ identity. By the above, $f \circ \pi = \pi$ is continuous, so f must be continuous. Reversing the roles of $\mathcal{T}_1$ and $\mathcal{T}_2$, this identity map must be a homeomorphism, so $\mathcal{T}_1 = \mathcal{T}_2$. (That is three times now!)

(e) Note that if U is open in G, then

$$f(UH/H) = f \circ \pi(U).$$

f is open when all those sets on the left are open, while $f \circ \pi$ is open when all those sets on the right are open.

(f) Note that if $B \in \mathcal{B}_e$, then $gH \in (\operatorname{int}(B))gH/H$, an open subset of BgH/H, so $gH \in \operatorname{int}(BgH/H)$. If $gH \in UH/H$, then $gH \subset UH$, so $g \in UH$, an open set. By Proposition 1.8, there exists $B \in \mathcal{B}_e$ with $Bg \subset UH$, so $BgH \subset UHH = UH$, so $BgH/H \subset UH/H$. These are the neighborhood base properties.

(g) Suppose $xH \neq yH$. Then $x \notin yH$. Now $\{x\}$ is compact, while yH is closed. Let $\mathcal{B}_e$ be any neighborhood base at e consisting of open sets. Then by Theorem 1.14 applied to the opposite group G^{op} (see p. 8), there exists $B \in \mathcal{B}_e$ for which $B\{x\} \cap ByH = \emptyset$. Now ByH is a union of left cosets of H, which are equivalence classes under "$u \sim v$ when $u^{-1}v \in H$," so nothing in $B\{x\} = Bx$ is equivalent to anything in By.

Hence $ByH \cap BxH = \emptyset$. Hence ByH/H is an open neighborhood of yH which is disjoint from the open neighborhood BxH/H of xH.

(i) It helps to do this before (h). Consider the natural bijection

$$\Phi : \left(\prod_{i \in \mathscr{I}} G_i\right) \Big/ \left(\prod_{i \notin \mathscr{I}} H_i\right) \to \prod_{i \in \mathscr{I}} (G_i/H_i).$$

It suffices to show that Φ is continuous and open. In view of parts (c) and (e), it suffices to show that

$$\Phi : \prod_{i \in \mathscr{I}} G_i \to \prod_{i \in \mathscr{I}} G_i/H_i$$

$$\Phi(\langle g_i \rangle) = \langle g_i H_i \rangle$$

is continuous and open. It is continuous by Theorem 1.18: The i_0 coordinate function of Φ is the composite $\Pi G_i \to G_{i_0} \to G_{i_0}/H_{i_0}$, which is continuous for every i_0. To verify that Φ is open takes a bit more work.

Suppose $\mathscr{F}$ is a finite subset of $\mathscr{I}$, and U_i is open in G_i when $i \in \mathscr{F}$. Then clearly Φ maps

$$\prod_{i \in \mathscr{I}} \left\{ \begin{array}{l} G_i \text{ if } i \notin \mathscr{F} \\ U_i \text{ if } i \in \mathscr{F} \end{array} \right\} \text{ to } \prod_{i \in \mathscr{I}} \left\{ \begin{array}{l} G_i/H_i \text{ if } i \notin \mathscr{F} \\ U_i H_i/H_i \text{ if } i \in \mathscr{F} \end{array} \right\}.$$

The point is that this is onto: Suppose $\langle g_i H_i \rangle$ satisfies $g_i H_i \in U_i H_i/H_i$ when $i \in \mathscr{F}$. Then $g_i H_i \subset U_i H_i$ when $i \in \mathscr{F}$. Choose $g_i' \in U_i$ with $g_i H_i = g_i' H_i$, when $i \in \mathscr{F}$. Set $g_i' = g_i$ when $i \notin \mathscr{F}$. Then $\Phi(\langle g_i' \rangle) = \langle g_i' H_i \rangle = \langle g_i H_i \rangle$.

So: Φ maps a base element B of the topology on $\prod G_i$ to a base element for the topology on $\prod(G_i/H_i)$. In particular, $\Phi(B)$ is open. But any open subset is a union of such base elements, and

$$\Phi(\cup B_\alpha) = \cup \Phi(B_\alpha) \text{ is open.}$$

(h) Inversion is continuous from G/H to G/H by part (c): The composite

$$G \xrightarrow{\quad \pi \quad} G/H \xrightarrow{\quad \text{inversion} \quad} G/H$$

equals

$$G \xrightarrow{\quad \text{inversion} \quad} G \xrightarrow{\quad \pi \quad} G/H$$

which is continuous.

Multiplication is continuous via the "diagram":

$$G \times G \twoheadrightarrow (G \times G)/(H \times H)$$

$$\wr\wr \xleftarrow{\makebox[3cm]{}} \text{part (i)}$$

$$(G/H) \times (G/H) \longrightarrow G/H.$$

By part (c), applied to $(G \times G)/(H \times H)$, it suffices to show that $G \times G \to G/H$, $(x, y) \mapsto xyH$, is continuous. But this, too, is the continuous composite

$$G \times G \xrightarrow{\quad \text{multiplication} \quad} G \xrightarrow{\quad \pi \quad} G/H.$$

□

One thing is worth noting regarding part (g): If G/H is just T_1, then the coset $H = eH$ is closed in G/H, so that $H = \pi^{-1}(H)$ is closed. Hence T_1 and T_2 are equivalent for quotient spaces of topological groups.

Part (g) does show the importance of closed subgroups. The following two results take care of closures and identification of closed subgroups.

Proposition 1.24. *Suppose G is a topological group, and H is a subgroup. Then H^- is also a subgroup. If H is normal, then so is H^-.*

Proof. H^- is closed under multiplication, since $(H^-) \cdot (H^-) \subset (H \cdot H)^- = H^-$ by Proposition 1.16. H^- is closed under the inversion map I since I is a homeomorphism and so preserves closures:

$$I(H^-) = (I(H))^- = H^-.$$

Finally, if H is normal, and $g \in G$, then setting $\tau_g(x) = gxg^{-1}$ we get that τ_g is a homeomorphism and again preserves closures:

$$\tau_g(H^-) = (\tau_g(H))^- = H^-.$$

□

Proposition 1.25. *Suppose G is a topological group with identity e, H is a subgroup of G, and C is a closed subset of G with $e \in \text{int}(C)$. Finally, suppose $H \cap C$ is closed. Then H is closed.*

Proof. The open neighborhoods of e form a neighborhood base at e, so there exist open neighborhoods W_1 and W_2 of e with $W_1 W_2 \subset \text{int}(C)$.

Suppose $p \in H^-$. Then $pW_1^{-1} \cap H \neq \emptyset$, so choose $q \in pW_1^{-1} \cap H$. Then $p^{-1}q \in W_1^{-1}$, so $q^{-1}p = (p^{-1}q)^{-1} \in W_1$. If U is open and $p \in U$, then $p \in pW_2 \cap U$, an open set, so $H \cap (pW_2 \cap U) \neq \emptyset$. That is, $(H \cap pW_2) \cap U \neq \emptyset$ whenever U is open and $p \in U$, so $p \in (H \cap pW_2)^-$. Hence $q^{-1}p \in q^{-1}(H \cap pW_2)^- = (q^{-1}H \cap q^{-1}pW_2)^-$ since left multiplication by q^{-1} is a homeomorphism. But $q^{-1}H = H$ since $q \in H$ and H is a subgroup, while $q^{-1}pW_2 \subset W_1 W_2 \subset \text{int}C \subset C$, so $q^{-1}p \in (H \cap C)^-$. But $H \cap C$ is closed, so $q^{-1}p \in H \cap C \subset H$. Hence $p = q \cdot q^{-1}p \in H$. $\qquad\square$

The results about quotients are typically applied to normal subgroups, with parts (c) and (e) of Theorem 1.23 being used when Y is also a topological group and $f : G/H \to Y$ is a homomorphism. Our final result for this section clarifies just what is required for homomorphisms in general.

Proposition 1.26. *Suppose G and $\tilde{G}$ are topological groups with identity elements e and $\tilde{e}$, respectively. Suppose $\mathscr{B}_e$ is a neighborhood base at e, and $\mathscr{B}_{\tilde{e}}$ is a neighborhood base at $\tilde{e}$. Finally, suppose $f : G \to \tilde{G}$ is a homomorphism.*

(a) f is continuous if, and only if, $e \in \text{int}(f^{-1}(\tilde{B}))$ for all $\tilde{B} \in \mathscr{B}_{\tilde{e}}$.

(b) f is an open map if, and only if, $\tilde{e} \in \text{int}(f(B))$ for all $B \in \mathscr{B}_e$.

Proof. (a) If f is continuous, then $f^{-1}(\text{int}(\tilde{B}))$ is open, and $e \in f^{-1}(\tilde{e}) \subset f^{-1}(\text{int}(\tilde{B})) \subset f^{-1}(\tilde{B})$, so $e \in \text{int} f^{-1}(\tilde{B})$. On the other hand, if $e \in \text{int}(f^{-1}(\tilde{B}))$ for all $\tilde{B} \in \mathscr{B}_{\tilde{e}}$, suppose U is open in $\tilde{G}$ and $g \in f^{-1}(U)$. Then $f(g) \in U$, so there exists $\tilde{B} \in \mathscr{B}_{\tilde{e}}$ with $f(g)\tilde{B} \subset U$. But f is a homomorphism: $f(gx) = f(g)f(x)$; taking $f(x) \in \tilde{B}$ (i.e. $x \in f^{-1}(\tilde{B})$), $f(gf^{-1}(\tilde{B})) \subset f(g)\tilde{B} \subset U$, that is $gf^{-1}(\tilde{B}) \subset f^{-1}(U)$. But $gf^{-1}(\tilde{B})$ is a neighborhood of g, so g is interior to $f^{-1}(U)$. Since g was arbitrary in $f^{-1}(U)$, $f^{-1}(U)$ is open.

(b) If f is open, then $\tilde{e} = f(e) \subset f(\text{int}(B))$, an open subset of $f(B)$, so $\tilde{e} \in \text{int} f(B)$. ($B \in \mathscr{B}_e$.) On the other hand, if $\tilde{e} \in \text{int}(f(B))$ for all $B \in \mathscr{B}_e$, suppose U is open in G. If $g \in U$, there exists $B \in \mathscr{B}_e$ with $gB \subset U$. Hence

$$f(g)f(B) = f(gB) \subset f(U).$$

But by assumption, $f(g)f(B)$ is a neighborhood of $f(g)$, so $f(g)$ is interior to $f(U)$. Since $g \in U$ was arbitrary, $f(U)$ is open. $\qquad\square$

1.5 Completeness

At long last, it is time to assume that our topological groups are Hausdorff. This is primarily motivated by Proposition 1.3(b): we want nets to have unique limits. We shall define the meaning of "Cauchy" here (it is slightly subtle), and define completeness as "Cauchy $\Rightarrow$ Convergent," as expected.

For metric spaces, a sequence $\langle x_n \rangle$ converges to x when the terms x_n get close to x. The sequence $\langle x_n \rangle$ is Cauchy when the terms x_n get close to each other. To illustrate what that translates into for topological groups, consider a convergent sequence $x_n \to x$ in a topological group G, with identity e. If B is a neighborhood of e, then for large n, $x_n \in xB$, that is $x^{-1}x_n \in B$.

Aha! That B is fixed; the group operation says (in a uniform sense) that x is close to y when $x^{-1}y \in B$.

Definition 1.27. Suppose G is a Hausdorff topological group, and $\langle x_\alpha \rangle$ is a net in G defined on a directed set D. $\langle x_\alpha \rangle$ is **left Cauchy** when the following happens: For every neighborhood B of the identity e of G, there exists an $\alpha_0 \in D$ such that

$$\forall \beta, \gamma \in D : \beta \succ \alpha_0 \text{ and } \gamma \succ \alpha_0 \Rightarrow x_\beta^{-1}x_\gamma \in B.$$

$\langle x_\alpha \rangle$ is **right Cauchy** when the following happens: For every neighborhood B of e, there exists an $\alpha_0 \in D$ such that

$$\forall \beta, \gamma \in D : \beta \succ \alpha_0 \text{ and } \gamma \succ \alpha_0 \Rightarrow x_\beta x_\gamma^{-1} \in B.$$

A subset A of G is called **complete** when each left Cauchy net in A is convergent to a point in A.

The terminology is slightly confusing. We should probably call A "left complete," but the terminology is usually applied to the whole group, where it does not matter. For the record, a left Cauchy sequence is simply a left Cauchy net defined on $(\mathbb{N}, \leq)$, and a subset A of G is sequentially complete when each left Cauchy sequence in A is convergent.

Proposition 1.28. *Suppose G is a complete Hausdorff topological group. Then each right Cauchy net converges.*

Proof. Suppose $\langle x_\alpha \rangle$ is a right Cauchy net defined on a directed set D. Set $y_\alpha = x_\alpha^{-1}$. Then for each neighborhood B of the identity e of G, there exists $\alpha_0 \in D$ such that for all $\beta, \gamma \in D$, $\beta \succ \alpha_0$ and $\gamma \succ \alpha_0 : x_\beta x_\gamma^{-1} \in B$, that is $y_\beta^{-1}y_\gamma \in B$. That is, $\langle y_\alpha \rangle$ is a left Cauchy net, so $\langle y_\alpha \rangle$ converges, say $y_\alpha \to y$. Then $x_\alpha = y_\alpha^{-1} \to y^{-1}$ since inversion is a homeomorphism. $\square$

There are some basic facts that need checking. The most basic is the following.

Proposition 1.29. *In any Hausdorff topological group, a convergent net is both left Cauchy and right Cauchy.*

Proof. Suppose $\langle x_\alpha \rangle$ is a convergent net defined on a directed set D, with $x_\alpha \to x$. Suppose B_1 is an open neighborhood of the identity e of G. Choose open neighborhoods B_2 and B_3 of e for which $B_2 B_3 \subset B_1$. Choose $\alpha_0 \in D$ so that $\beta \succ \alpha_0 \Rightarrow x_\beta \in x(B_2^{-1} \cap B_3)$. Then for $\beta, \gamma \succ \alpha_0$: $x_\gamma \in x(B_2^{-1} \cap B_3) \subset xB_3$, so $x^{-1}x_\gamma \in B_3$, while $x_\beta \in x(B_2^{-1} \cap B_3) \subset xB_2^{-1}$, so $x^{-1}x_\beta \in B_2^{-1}$, so $x_\beta^{-1}x = (x^{-1}x_\beta)^{-1} \in B_2$. Hence $x_\beta^{-1}x_\gamma = x_\beta^{-1}x \cdot x^{-1}x_\gamma \in B_2 B_3 \subset B_1$.

Right Cauchy is similar. Choose $\alpha_1 \in D$ so that $\beta \succ \alpha_1 \Rightarrow x_\beta \in (B_2 \cap B_3^{-1})x$. Then $\beta, \gamma \succ \alpha_1 \Rightarrow x_\beta \in (B_2 \cap B_3^{-1})x \subset B_2 x$, so $x_\beta x^{-1} \in B_2$, while $x_\gamma \in (B_2 \cap B_3^{-1})x \subset B_3^{-1}x$, so $x_\gamma x^{-1} \in B_3^{-1}$, so $xx_\gamma^{-1} = (x_\gamma x^{-1})^{-1} \in B_3$ and $x_\beta x_\gamma^{-1} = x_\beta x^{-1} \cdot x x_\gamma^{-1} \in B_2 B_3 \subset B_1$. □

This suggests that maybe we should have defined "complete" by requiring nets that are *both* left Cauchy and right Cauchy to be convergent. It turns out that while this is a possible approach, it is just not as fruitful as the one here. There do exist groups for which every right-Cauchy-and-left-Cauchy net converges, but which are not complete. See Chap. 7 of Dikranjan, Prodanov, and Stoyanov [11] for a discussion.

Example 2. Let $S(\mathbb{N})$ denote the set of bijections of $\mathbb{N}$ with itself, and $\mathscr{B}_e =$ all stabilizers of finite subsets of $\mathbb{N}$. As noted earlier, $S(\mathbb{N})$ is a topological group with $\mathscr{B}_e$ as a neighborhood base at the identity, e. Set $x_n = (0, 1, 2, \ldots, n)$. It is easy to see that $x_m^{-1}x_n$ fixes all integers $< \min(m, n)$, so $\langle x_n \rangle$ is left Cauchy. It is *not* right Cauchy, since $x_n x_m^{-1}$ sends 0 to m (when $n < m$) or $m + 1$ (when $n > m$). In fact, if you look at what x_n does "pointwise," it converges to the function sending n to $n + 1$, a function that is not onto. It follows that $S(\mathbb{N})$ cannot be imbedded in a complete group, since $\langle x_n \rangle$ would have to converge there, and so would have to be right Cauchy.

We have a few more results here for general nets. The first two are "expected" from our experience with completeness in metric spaces, and are contained in the following result.

Proposition 1.30. *Suppose G is a Hausdorff topological group, and suppose A is a complete subset of G. Then A is closed in G, and any closed subset of A is also complete.*

Proof. A is closed: Suppose $x \in A^-$. Then by Proposition 1.3(a), there is a net $\langle x_\alpha \rangle$ defined on a directed set D such that $x_\alpha \in A$ and $x_\alpha \to x$. But now $\langle x_\alpha \rangle$ is left Cauchy by Proposition 1.29, so x_α must converge to a point $y \in A$ by definition of "complete." Finally, $y = x$ by Proposition 1.3(b): limits of nets are unique since G is Hausdorff. Hence $x = y \in A$. Thus, all points of A^- are in A, so A is closed.

Suppose $B \subset A$, and B is closed. Let $\langle x_\alpha \rangle$ be a left Cauchy net in B, defined on a directed set D. Then $\langle x_\alpha \rangle$ is a left Cauchy net in A, so $x_\alpha \to x \in A$ since A is complete. But B is closed, so $x \in B$ by Proposition 1.3(a). □

We recall a definition for our next result. Suppose G is a Hausdorff topological group, and $\langle x_\alpha \rangle$ is a net in G defined on a directed set D. A point $x \in G$ is called a

cluster point of $\langle x_\alpha \rangle$ if, for each open set U with $x \in U$, and $\alpha \in D$: there exists $\beta \in D$ with $\beta \succ \alpha$ and $x_\beta \in U$.

Proposition 1.31. *Suppose G is a Hausdorff topological group, and suppose $\langle x_\alpha \rangle$ is a left Cauchy net with a cluster point x. Then $x_\alpha \to x$.*

Proof. Suppose $\langle x_\alpha \rangle$ is a left Cauchy net defined on a directed set D, with a cluster point x. Suppose U is open, with $x \in U$. Choose an open neighborhood B_1 of the identity e in G for which $xB_1 \subset U$. Choose open neighborhoods B_2 and B_3 of e for which $B_2 B_3 \subset B_1^{-1}$. Choose $\alpha_0 \in D$ for which $\beta, \gamma \in D$, $\beta \succ \alpha_0, \gamma \succ \alpha_0 \Rightarrow x_\beta^{-1} x_\gamma \in B_2$. Choose $\gamma \in D$, $\gamma \succ \alpha_0$ for which $x_\gamma \in xB_3^{-1}$, an open set containing x. Then $x^{-1} x_\gamma \in B_3^{-1}$, so $x_\gamma^{-1} x = (x^{-1} x_\gamma)^{-1} \in B_3$, so $x_\beta^{-1} x = x_\beta^{-1} x_\gamma \cdot x_\gamma^{-1} x \in B_2 B_3 \subset B_1^{-1}$ whenever $\beta \succ \alpha_0$. But now $x^{-1} x_\beta = (x_\beta^{-1} x)^{-1} \in B_1$, so $x_\beta \in xB_1 \subset U$ when $\beta \succ \alpha_0$. $\square$

Corollary 1.32. *Suppose G is a Hausdorff topological group, and K is a compact subset of G. Then K is complete.*

Proof. Suppose $\langle x_\alpha \rangle$ is a left Cauchy net in K. Then $\langle x_\alpha \rangle$ has a cluster point $x \in K$ by Proposition 1.5, and $\lim x_\alpha = x$ by Proposition 1.31. $\square$

Corollary 1.33. *Suppose G is a Hausdorff topological group and suppose $\langle x_\alpha : \alpha \in D \rangle$ is a left Cauchy net in G. Suppose D' is cofinal in D, and suppose $\lim_{D'} x_\alpha = x$. Then $\lim_D x_\alpha = x$.*

Proof. In view of Proposition 1.31, it suffices to show that x is a cluster point of $\langle x_\alpha : \alpha \in D \rangle$. But if U is an open neighborhood of x, then there exists $\beta \in D'$ such that $\gamma \succ \beta$ and $\gamma \in D' \Rightarrow x_\gamma \in U$. Now suppose $\alpha \in D$. There exists $\gamma \in D$ s.t. $\gamma \succ \alpha$ and $\gamma \succ \beta$ since D is directed, and there exists $\delta \in D'$ s.t. $\delta \succ \gamma$ since D' is cofinal. But now $\delta \succ \gamma \succ \beta$, so $x_\delta \in U$, and $\delta \succ \gamma \succ \alpha$, so $\delta \succ \alpha$. That is, x *is* a cluster point of $\langle x_\alpha : \alpha \in D \rangle$. $\square$

There is one consequence of the above: Our definition of "complete" is not *too* loose. If it were, we would not be able to prove Proposition 1.31 or Corollary 1.32.

These are the basics for completeness. A lot more can be said when our Hausdorff topological group is also first countable, enough for a section all its own.

1.6 Completeness and First Countability

One fact we have gotten used to when working with metric spaces is that sequences are "enough." Does that apply to completeness as well when groups are first countable? Yes!

Theorem 1.34. *Suppose G is a first countable Hausdorff topological group, and suppose G is sequentially complete. Then G is complete.*

Proof. Suppose G is a first countable Hausdorff topological group in which each left Cauchy sequence converges. Suppose $\langle x_\alpha \rangle$ is a left Cauchy net defined on a directed set D. Choose any neighborhood base at the identity e in G which is in accord with Theorem 1.13: $\mathscr{B}_e = \{B_1, B_2, \ldots\}$, with $B_j = B_j^{-1} \supset B_{j+1}^2$ for $j = 1, 2, \ldots$. Define α_n recursively as follows. Choose $\alpha_1 = \alpha_1'$ so that $\beta, \gamma \succ \alpha_1 \Rightarrow x_\beta^{-1} x_\gamma \in B_1$. Given α_n, choose α_{n+1}' so that $\beta, \gamma \succ \alpha_{n+1}' \Rightarrow x_\beta^{-1} x_\gamma \in B_{n+1}$, and choose α_{n+1} so that $\alpha_{n+1} \succ \alpha_n$ and $\alpha_{n+1} \succ \alpha_{n+1}'$. Consider the sequence $y_n = x_{\alpha_n}$. If $m > n$, then $\alpha_m \succ \alpha_n \succ \alpha_n'$, so $x_{\alpha_m}^{-1} x_{\alpha_n} \in B_n$. That is, $y_m^{-1} y_n \in B_n$. Also, $y_n^{-1} y_m = (y_m^{-1} y_n)^{-1} \in B_n^{-1} = B_n$.

Suppose U is open, with $e \in U$. Choose N with $B_N \subset U$. If $m, n \geq N$, then $y_m^{-1} y_n \in B_{\min(m,n)} \subset B_N \subset U$. So: $\langle y_n \rangle$ is a left Cauchy sequence, so $y_n \to y$ for some $y \in G$ since we are assuming that all left Cauchy sequences converge. The claim is that $x_\alpha \to y$.

Suppose U is open, and $y \in U$. Choose N with $y B_N \subset U$. Choose N' so that $n \geq N' \Rightarrow y_n \in y B_{N+1}$. Set $n_0 = \max(N', N+1)$, and suppose $\beta \succ \alpha_{n_0}$. Then $x_{\alpha_{n_0}}^{-1} x_\beta \in B_{n_0} \subset B_{N+1}$, since $\beta \succ \alpha_{n_0} \succ \alpha_{n_0}'$. That is, $y_{n_0}^{-1} x_\beta \in B_{N+1}$. But also, $y_{n_0} \in y B_{N+1}$ since $n_0 \geq N'$. Hence $y^{-1} y_{n_0} \in B_{N+1}$. Hence $y^{-1} x_\beta = y^{-1} y_{n_0} \cdot y_{n_0}^{-1} x_\beta \in B_{N+1}^2 \subset B_N$, so $x_\beta \in y B_N \subset U$. In a nutshell,

$$\beta \succ \alpha_{n_0} \Rightarrow x_\beta \in U.$$

That is convergence.   □

Theorem 1.34 illustrates why the use of sequences is sufficient; it does not explain why it matters. Sequences allow parametrization by $\mathbb{N}$, the natural numbers, and $\mathbb{N}$ is not just directed, it is well-ordered. This allows definition by recursion, which has already occurred in Theorem 1.13. Furthermore, the recursive definitions only require finite intermediate constructions. This kind of thing also occurs in situations devoid of sequences; the proof that a regular Lindelöf space is normal is particularly blatant, cf. Kelley [20].

To proceed further, we need to fix some neighborhood base $\mathscr{B}_e = \{B_1, B_2, \ldots\}$ with $B_j = B_j^{-1} \supset B_{j+1}^2$. Observe that $B_j B_{j+1} \cdots B_{j+n} \subset B_{j-1}$ for $j \geq 2$, by induction on n: The $n = 1$ case follows from $B_{j+1} = e B_{j+1} \subset B_{j+1}^2 \subset B_j$, so $B_j B_{j+1} \subset B_j^2 \subset B_{j-1}$, while the statement

$$\forall \, j \geq 2 : B_j B_{j+1} \cdots B_{j+n} \subset B_{j-1}$$

holds by induction on n. By the induction hypothesis:

$$B_j B_{j+1} \cdots B_{j+(n+1)} = B_j (B_{j+1} \cdots B_{(j+1)+n}) \subset B_j B_j \subset B_{j-1}.$$

Now suppose we are given some $x_n \in B_n$ for all n. Set

$$y_n = x_1 x_2 \cdots x_n = \prod_{j=1}^{n} x_j.$$

The order matters here! Note that if $m > n$, then

$$y_n^{-1} y_m = x_n^{-1} x_{n-1}^{-1} \cdots x_1^{-1} x_1 x_2 \cdots x_n \cdots x_m$$

$$= x_{n+1} \cdots x_m \in B_{n+1} B_{n+2} \cdots B_m \subset B_n.$$

Also,

$$y_m^{-1} y_n = (y_n^{-1} y_m)^{-1} \in B_n^{-1} = B_n.$$

If U is open and $e \in U$, then one can choose N with $B_N \subset U$. If $n, m \geq N$, then

$$y_m^{-1} y_n \in B_{\min(m,n)} \subset B_N \subset U.$$

That is, $\langle y_n \rangle$ is a left Cauchy sequence, so if G is complete, then all such products must converge. We need a name for all this. Call such an infinite product

$$\prod x_j = \text{``} \lim_{n \to \infty} \text{''} x_1 x_2 \cdots x_n$$

a "$\langle B_n \rangle$-compatible infinite product."

Theorem 1.35. *Suppose G is a first countable Hausdorff topological group. Suppose $\mathscr{B}_e = \{B_1, B_2, \ldots\}$ is a neighborhood base at the identity e in G for which $B_j = B_j^{-1} \supset B_{j+1}^2$ for all G. Then G is complete if, and only if, every $\langle B_n \rangle$-compatible infinite product is convergent.*

Proof. The "only if" part is done. For the "if" part, suppose every $\langle B_n \rangle$-compatible infinite product converges, and suppose $\langle g_m \rangle$ is a Cauchy sequence. For each n, choose M_n so that $k, l \geq M_n \Rightarrow g_k^{-1} g_l \in B_n$. Choose m_n recursively so that $m_n \geq M_n$ and $m_n > m_{n-1}$, with $m_1 = M_1$. Then $m_{n+1} > m_n \geq M_n$, so $x_n = g_{m_n}^{-1} g_{m_{n+1}} \in B_n$ for all n. By assumption, the infinite product $\prod x_n$ converges. Set $y_n = x_1 x_2 \cdots x_n \to y$. Then

$$y_n = g_{m_1}^{-1} g_{m_2} g_{m_2}^{-1} g_{m_3} \cdots g_{m_n}^{-1} g_{m_{n+1}} = g_{m_1}^{-1} g_{m_{n+1}}$$

so $y_n = g_{m_1}^{-1} g_{m_{n+1}} \to y$, giving $g_{m_{n+1}} \to g_{m_1} y$. Since the subsequence $\langle g_{m_{n+1}} \rangle$ converges to $g_{m_1} y$, $\langle g_m \rangle$ converges by Corollary 1.33. Hence G is complete by Theorem 1.34. $\square$

Corollary 1.36. *Suppose G is a first countable Hausdorff topological group, and H is a closed subgroup. Then: If G is complete, then so are H and G/H.*

Proof. H is complete by Proposition 1.30. As for G/H, suppose $\mathscr{B}_e = \{B_1, B_2, B_3, \ldots\}$ is a neighborhood base at the identity e of G, with $B_j = B_j^{-1} \supset B_{j+1}^2$ for all j. Then $\{B_j gH/H \; : \; B_j \in \mathscr{B}_e\}$ is a (countable) neighborhood base at $gH \in G/H$ by Theorem 1.23(f). In particular, G/H is first countable. G/H is also Hausdorff by Theorem 1.23(g). Finally, $(B_j H/H) = (B_j^{-1} H/H) = (B_j H/H)^{-1} \supset (B_{j+1}^2 H/H) = (B_{j+1} H/H)^2$ by how inverses and products are computed in G/H, so by Theorem 1.35, it suffices to show that any $\langle B_j H/H \rangle$-compatible infinite product in G/H is convergent.

Suppose $x_j H \in B_j H/H$, with $x_j \in B_j$. Then $\prod x_j$ converges, that is $y_n = x_1 x_2 \cdots x_n \to y$. But now $y_n H = x_1 H x_2 H \cdots x_n H \to yH$ since the natural map $\pi : G \to G/H$ is continuous [Theorem 1.23(b)].   □

By the way, Hausdorff quotients of complete Hausdorff topological groups that are not first countable need not be complete. An example of this is constructed in Exercise 32 of Chap. 5.

A final note concerning first countable Hausdorff topological groups. All such groups are metrizable, with a metric that is left invariant; see Hewitt and Ross [17, p. 70, Theorem 8.3]. In any such group, a Cauchy sequence in the metric is left Cauchy in the topological group (see Exercise 16). Consider this more evidence that the "right" notion of completeness simply requires that all left Cauchy nets be convergent.

Exercises

1. (Net practice) Suppose X is a topological space, and $\langle x_\alpha : \alpha \in D \rangle$ is a net in X. Show that any limit of $\langle x_\alpha \rangle$ is a cluster point of $\langle x_\alpha \rangle$.
2. Suppose X is a topological space, and $\langle x_\alpha : \alpha \in D \rangle$ is a net in X. Suppose D' is a directed set, and suppose $\varphi : D' \to D$ defines a subnet $\langle x_{\varphi(\alpha')} : \alpha' \in D' \rangle$ of $\langle x_\alpha : \alpha \in D \rangle$.

   (a) Show that if $\lim_D x_\alpha = x$, then $\lim_{D'} x_{\varphi(\alpha')} = x$.
   (b) Show that if y is a cluster point of $\langle x_{\varphi(\alpha')} : \alpha' \in D' \rangle$, then y is a cluster point of $\langle x_\alpha : \alpha \in D \rangle$.

3. Suppose X is a topological space, $\langle x_\alpha : \alpha \in D \rangle$ is a net in X, and y is a cluster point of $\langle x_\alpha : \alpha \in D \rangle$. This problem is concerned with constructing a subnet of $\langle x_\alpha : \alpha \in D \rangle$ that converges to y. Let $\mathscr{B}_y$ denote a neighborhood base at y, and set $D' = D \times \mathscr{B}_y$. Declare that $(\alpha, B) \geq (\beta, B')$ when $\alpha \succ \beta$ and $B \subset B'$. Finally, given $(\alpha, B) \in D'$, define $\varphi : D' \to D$ as follows. Since y is a cluster point of the original net, choose $\gamma = \varphi(\alpha, B)$ so that $\gamma \succ \alpha$ (in D) and $x_\gamma \in \text{int}(B)$.

   Note: To show that this is a subnet converging to y, there are three things to do: First, show that D' is a directed set. Second, show that the map φ produces a subnet; the fact that $\varphi(\alpha, B) \succ \alpha$ in D will help here. Finally show that

$\lim_{D'} x_{\varphi(\alpha, B)} = y$. Here, use the fact that $x_{\varphi(\alpha, B)} \in B$. You should also see why it is crucial that D be nonempty: Given an open set $U \subset X$ with $y \in U$, fix $B_0 \in \mathscr{B}_y$ with $B_0 \subset U$, *and any* $\alpha_0 \in D$. If $(\beta, B) \succ (\alpha_0, B_0) \ldots$.

4. Examine the preceding three exercises, and prove the following: Given a net $\langle x_\alpha : \alpha \in D \rangle$ in a topological space X, a point $y \in X$ is a cluster point of $\langle x_\alpha : \alpha \in D \rangle$ if and only if x is a limit of a subnet of $\langle x_\alpha : \alpha \in D \rangle$.

5. Suppose X is a topological space that is not compact. Suppose $\mathscr{C}$ is an open cover of X that has no finite subcover. The point here is to use $\mathscr{C}$ to construct a net in X which has no cluster points. The main obstruction is comprehensibility. The directed set D is the collection of all finite subsets of $\mathscr{C}$, partially ordered by set inclusion. If $\alpha \in D$, say $\alpha = \{U_1, U_2, \ldots, U_n\}$, then α does not cover X; choose any

$$x_\alpha \in X - \bigcup_{k=1}^{n} U_k.$$

Show that if $p \in X$, and $U \in \mathscr{C}$, $p \in U$, then $\alpha \supset \{U\}$ implies that $x_\alpha \notin U$. Use this to show that p is not a cluster point of $\langle x_\alpha \rangle$.

Now put it all together; prove the equivalence of the following three statements, for any topological space X:

(i) X is compact.
(ii) Every net in X has a cluster point.
(iii) Every net in X has a convergent subnet.

6. Suppose X is a set, and suppose for each $x \in X$ there is assigned a nonempty family $\mathscr{B}_x$ of subsets of X, subject to:

(a) For all $B \in \mathscr{B}_x : x \in B$.
(b) For all $B_1, B_2 \in \mathscr{B}_x$ there exists $B_3 \in \mathscr{B}_x$ with $B_3 \subset B_1 \cap B_2$.

Set

$$\mathscr{T} = \{U \subset X : \forall x \in U \, \exists B \in \mathscr{B}_x \text{ with } B \subset U\}.$$

Show that $\mathscr{T}$ is a topology. Show that the following are equivalent:

(i) For all $x \in X$, $\mathscr{B}_x$ is a neighborhood base at x for the topology $\mathscr{T}$.
(ii) For all $A \subset X$, $\text{int}(A) = \{x \in A : \text{there exists } B \in \mathscr{B}_x \text{ with } B \subset A\}$.
(iii) For all $x \in X$, for all $B_1 \in \mathscr{B}_x$, there exists $B_2 \in \mathscr{B}_x$ such that if $y \in B_2$, then there exists $B_3 \in \mathscr{B}_y$ with $B_3 \subset B_1$.

Note and Hint: Condition (iii) is meant to convey that the elements $y \in B_2$ will wind up interior to B_1.

7. Suppose X is a topological space. Show that the following are equivalent:

   (i) Each point of X has a neighborhood base consisting of open connected sets.
   (ii) Each point of X has a neighborhood base consisting of connected sets.
   (iii) Components of open subsets of X are open.

8. (A general construction of non-Hausdorff [indeed, non-T_1] spaces, useful for counterexamples galore.) Suppose X and Y are two disjoint topological spaces. (If they are not disjoint, replace one with a homeomorphic copy that is disjoint from the other.) Define $X \diagup Y$ to be $X \cup Y$, with the following "topology": The open subsets of $X \diagup Y$ are one of the following two types:

   (a) Open subsets of Y, or
   (b) *All* of Y, unioned with an open subset of X.

   (The pictorial idea behind the notation is that X is ramped up over Y, where its nonempty open subsets "leak" and fill up Y.)

   (a) Show that this is a topology.
   (b) Show that this topology is T_0 if both X and Y are T_0.
   (c) Show that this topology is never T_1 if both X and Y are nonempty.
   (d) Show that if K is compact and nonempty in X, and $A \subset Y$, then $K \cup A$ is compact in $X \diagup Y$.

9. Consider the topology constructed in Exercise 8, in conjunction with the three conditions appearing in Proposition 1.6:

   (i) Each point has a neighborhood base consisting of compact sets.
   (ii) Each point has an open neighborhood with compact closure.
   (iii) Each point has a compact neighborhood.

   As noted in Sect. 1.1, (i) $\Rightarrow$ (iii) $\Leftarrow$ (ii) always.

   (a) Set $X = \{\sqrt{2}\}$, $Y = \mathbb{Q}$, usual topologies. Show that $X \diagup Y$ satisfies (ii) and (iii) but not (i).
   (b) Set $X = \mathbb{Z}$ with the usual topology, and $Y = (0, 1)$ with the cofinite topology. Show that $X \diagup Y$ satisfies (i) and (iii) but not (ii).

10. (Yet another counterexample.) Set $X = \mathbb{Z}$ with the cofinite topology, and $Y = (0, 1)$ with the discrete topology. set $C_n = \{n, n+1, \ldots\} \cup Y$, so that $C_1 \supset C_2 \supset \cdots$. Show that, in $X \diagup Y$ (Exercise 8), each C_n is compact and connected, while $\cap C_n$ is neither.

11. Suppose G is a topological group, and $A, B \subset G$. Set $[A, B] = \{aba^{-1}b^{-1} : a \in A, b \in B\}$. Show that $[A^-, B^-] \subset [A, B]^-$.

12. Suppose G is a topological group. Define the derived series as usual: $G^0 = G$, and $G^{n+1} =$ subgroup generated by $[G^n, G^n]$.

Define $G^{[0]} = G$, and $G^{[n+1]} = $ closure of the subgroup generated by $[G^{[n]}, G^{[n]}]$. (Think of this as the topologically derived series.) Show that $(G^n)^- = G^{[n]}$ for all n. Hence show that $G^{[n]} = \{$identity$\}$ for some n if, and only if, G is solvable and Hausdorff.

13. Suppose G is a topological group, and H is a subgroup. Let $\pi : G \to G/H$ denote the canonical map. Show that the quotient topology on G/H is the unique topology $\mathscr{T}$ on G/H with the following property: If $f : G/H \to Y$ is a function, where Y is a topological space, then f is an open map if, and only if, $f \circ \pi$ is an open map.

14. Suppose G is a Hausdorff topological group that is also locally compact. Show that G is complete.

Suggestion. Let C be a compact neighborhood of the identity e of G. If $\langle x_\alpha \rangle$ is a left Cauchy net defined on a directed set D, choose α_0 so that $\beta, \gamma \succ \alpha_0 \Rightarrow x_\beta^{-1} x_\gamma \in C$.

Now look at $\langle x_{\alpha_0}^{-1} x_\gamma : \gamma \succ \alpha_0 \rangle$, a net on a cofinal subset of D.

15. Suppose G is a Hausdorff topological group.

(a) Show that all centralizers of subsets are closed. In particular, the center is closed.

(b) Show that the ascending central series $[Z_1(G) = $ center; $Z_{n+1}(G)$ =pullback in G of the center of $G/Z_n(G)]$ consists of closed subgroups. (So "Topologically nilpotent"= nilpotent.)

(c) Show that the normalizer of a closed subgroup is closed. *Hint:* There is a subtlety here. If H is the subgroup, then $A = \{g \in G : gHg^{-1} \subset H\}$ is *not* the normalizer in general, but $A \cap A^{-1}$ is. Note that $A = \bigcap_{h \in H} \{g \in G : ghg^{-1} \in H\}$.

Remark: For part (a), it may be helpful (depending on your approach) to prove the following. If X and Y are topological spaces, with Y Hausdorff, and $f, g : X \to Y$ are two continuous functions, then the equalizer $\{x \in X : f(x) = g(x)\}$ is closed in X.

16. Suppose G is a topological group for which the topology comes from a left invariant metric d: That is, $d(xy, xz) = d(y, z)$ for all $x, y, z \in G$. Show that a sequence $\langle x_n \rangle$ in G is left Cauchy in G as a topological group if, and only if, $\langle x_n \rangle$ is Cauchy under the metric d.

17. Suppose G is a commutative topological group, and suppose d is a metric on G which is (left) invariant; that is, $d(xy, xz) = d(y, z)$ for all $x, y, z \in G$. Show that, in the metric topology, G is a topological group. (It will help to show that $x \mapsto x^{-1}$ is an isometry.)

18. Suppose G is a topological group, and suppose H and K are subgroups, with $H \subset K \subset G$, so that $K/H \subset G/H$. Show that K is closed in G if and only if K/H is closed in G/H.

19. Suppose G and H are two Hausdorff topological groups, and $f : G \to H$ is a homomorphism.

(a) Suppose f is continuous, and $\langle x_\alpha \rangle$ is a left Cauchy net in G. Show that $\langle f(x_\alpha) \rangle$ is a left Cauchy net in H.

(b) Suppose the graph of f,

$$\Gamma(f) = \{(x, y) \in G \times H : y = f(x)\},$$

is closed in $G \times H$; and suppose $\tilde{H}$ is a complete subgroup of H. Set $\tilde{G} = f^{-1}(\tilde{H})$, and suppose the restriction

$$f \bigg|_{\tilde{G}} : \tilde{G} \to \tilde{H}$$

is continuous. Show that $\tilde{G}$ is closed in G. (Consult Proposition A.2 in Appendix A.)

20. Suppose G is a topological group, and H is a subgroup. Suppose int(H) is nonempty. Show that H is both open and closed, and the quotient topology on G/H is the discrete topology.

Chapter 2
Topological Vector Spaces

2.1 Generalities

A topological vector space X over $\mathbb{R}$ or $\mathbb{C}$ is a vector space, which is also a topological space, in which the vector space operations are continuous. Letting $\mathbb{F}$ denote the field $\mathbb{R}$ or $\mathbb{C}$ we require the maps

$$\text{addition}: X \times X \to X,$$
$$(x, y) \mapsto x + y; \text{ and}$$

$$\text{scalar multiplication}: \mathbb{F} \times X \to X$$
$$(r, x) \mapsto rx$$

to be continuous. Note that scalar multiplication is jointly continuous and hence is separately continuous. In particular, multiplication by -1 is continuous. This is just "inversion" in the additive group, so *every topological vector space is a topological group*. In particular, the theorems from Chap. 1 are all available for topological vector spaces. Banach spaces, with the norm topology, are topological vector spaces.

NONEXAMPLES: (1) If V is an infinite-dimensional vector space over $\mathbb{F}$, and $\mathscr{B}_0 =$ all subspaces of finite codimension, then, as noted earlier, $(V, +)$ becomes a topological group with $\mathscr{B}_0$ as a neighborhood base at 0. It is not a topological vector space, since scalar multiplication is not even separately continuous. (It is not continuous in r.) (2) $\mathbb{R}^2$, with the "Washington metric," where the distance between x and y is $\|x - y\|$ if x and y are colinear and is $\|x\| + \|y\|$ if not. (Roughly speaking, the distance between x and y is the distance you must travel to get from x to y when you are only allowed to move radially. It gets its name from the street plan of Washington, D.C.) Scalar multiplication is continuous, but addition is not even separately continuous in this metric.

M.S. Osborne, *Locally Convex Spaces*, Graduate Texts in Mathematics 269,
DOI 10.1007/978-3-319-02045-7_2, © Springer International Publishing Switzerland 2014

It is evident that the definition really only requires $\mathbb{F}$ to be a "topological field," and some things can be done in this context. However, convexity arises early, and this will require X to be a vector space over $\mathbb{R}$. This, in turn, will require our field $\mathbb{F}$ to be an extension field of $\mathbb{R}$. Since it will also be handy for $\mathbb{F}$ to be locally compact, transcendental extensions are out, and we are left with $\mathbb{R}$ or $\mathbb{C}$. Since these give the most useful examples, this is not a strong restriction. Nevertheless, there are some cases (not discussed in this book) where authors look seriously at topological vector spaces over p-adic fields, for example, Escassut [14] or Van Rooij [38].

The Hausdorff condition is not assumed here for topological vector spaces. However, we shall soon restrict attention to locally convex Hausdorff topological vector spaces; the Hausdorff condition will be required to make any real headway exploiting the "locally convex" condition. Of course, the adverb "locally" in "locally convex" means exactly what an adverb should, as described in Sect. 1.1.

A set A in a vector space over $\mathbb{R}$ is *convex* if:

$$\forall\, x, y \in A, \forall\, t \in [0, 1] : tx + (1 - t)y \in A.$$

A topological vector space is locally convex if every point has a neighborhood base consisting of convex sets. A locally convex topological vector space will be called a locally convex space (abbreviated "LCS" or "l.c.s." in the literature) for short. Convexity is a major topic and is worthy of its own section.

The following example (actually a whole class of examples) follows a routine similar to the one for norms. In many cases, these spaces will not be locally convex and will eventually serve that role. For now, they are simply examples of topological vector spaces constructed using a metric. A general neighborhood-base construction for locally convex spaces will be provided later, in Sect. 3.1.

Example 3. L^p-spaces, $0 < p < 1$. Let $(X, \mathscr{B}, \mu)$ denote a measure space and, as usual, declare two measurable functions to be equivalent when they are equal a.e. Letting $[f]$ denote the function class of f, set

$$L^p(\mu) = \left\{ [f] : \int |f|^p d\mu < \infty \right\}, \text{ and}$$

$$\rho([f]) = \int |f|^p d\mu.$$

The straightforward inequality $(s + t)^p \leq s^p + t^p$ when $s, t \geq 0$:

$$(s + t)^p - s^p = \int_s^{s+t} px^{p-1} dx = \int_0^t p(x + s)^{p-1} dx$$

$$\leq \int_0^t px^{p-1} dx = t^p,$$

yields both the fact that $L^p(\mu)$ is a vector space as well as the inequality $\rho([f] + [g]) \leq \rho([f]) + \rho([g])$; we also get that $\rho([f]) = 0 \Leftrightarrow [f] = [0]$ and $\rho(c[f]) = |c|^p \rho([f])$. When $p = 1$, these are the conditions defining a norm, and whether

$p = 1$ or $p \in (0, 1)$, they suffice to manufacture a topological vector space using the metric $d([f], [g]) = \rho([f] - [g])$. Given $[f]$, $[g]$, and $[h]$,

$$d([f], [g]) = \rho([f] - [h]) = \rho([f] - [g] + [g] - [h])$$
$$\leq \rho([f] - [g]) + \rho([g] - [h])$$
$$= d([f], [g]) + d([g], [h]);$$
$$d([g], [f]) = \rho([g] - [f]) = \rho(-1([f] - [g]))$$
$$= |-1|^p \rho([f] - [g]) = d([f], [g]),$$

so d is a metric, while

$$d([f] + [g], [f] + [h]) = \rho([f] + [g] - [f] - [h])$$
$$= \rho([g] - [h]) = d([g], [h]),$$

so d is translation invariant. By Exercise 17 in Chap. 1, $L^p([\mu])$ is at least a topological group with this metric, while joint continuity of scalar multiplication is easily checked using sequences: If $c_n \to c$ and $[f_n] \to [f]$, then

$$\rho(c_n[f_n] - c[f]) = \rho(c_n[f_n] - c_n[f] + c_n[f] - c[f])$$
$$\leq \rho(c_n[f_n] - c_n[f]) + \rho(c_n[f] - c[f])$$
$$= \rho(c_n([f_n] - [f])) + \rho((c_n - c)[f])$$
$$= |c_n|^p d([f_n], [f]) + |c_n - c|^p \rho([f])$$
$$\to |c|^p \cdot 0 + 0 \cdot \rho([f]) = 0.$$

Topological vector spaces can arise from metrics, as these spaces (and normed spaces as well) do. They can arise in much more complicated ways as well. We close this section with the standard constructions borrowed from the underlying algebra.

Theorem 2.1. *Suppose $\langle X_i, i \in \mathscr{I} \rangle$ is a family of topological vector spaces. Then with the product topology, $\prod X_i$ is a topological vector space.*

Proof. $\prod X_i$ is a topological group by Corollary 1.20; the only issue is the joint continuity of scalar multiplication. Letting $\mathbb{F}$ denote the base field, note that the "diagonal" map

$$\mathbb{F} \to \prod_{i \in \mathscr{I}} \mathbb{F}$$

is continuous, since each component function is the identity function (Theorem 1.18). Hence the maps

$$\mathbb{F} \times \prod_{i \in \mathscr{I}} X_i \to \left(\prod_{i \in \mathscr{I}} \mathbb{F} \right) \times \left(\prod_{i \in \mathscr{I}} X_i \right)$$

$$\approx \prod_{i \in \mathscr{I}} (\mathbb{F} \times X_i) \to \prod_{i \in \mathscr{I}} X_i$$

are continuous: The intermediate isomorphism is from Corollary 1.19, and the last arrow is continuous because its coordinate functions are

$$\pi_{j_0} : \prod_{i \in \mathscr{I}} (\mathbb{F} \times X_i) \to \mathbb{F} \times X_{j_0} \xrightarrow{\text{mult.}} X_{j_0},$$

which are continuous. □

Theorem 2.2. *Suppose X is a topological vector space, and Y is a vector subspace. Then with the induced topology, Y is a topological vector space. Also, with the quotient topology, X/Y is a topological vector space that is Hausdorff if, and only if, Y is closed.*

Proof. Again, the only issue is scalar multiplication: Y is a topological group by Corollary 1.22, while X/Y is a topological group by Theorem 1.23(h); Theorem 1.23(g) addresses the Hausdorff condition. Letting $\mathbb{F}$ denote the base field,

$$\mathbb{F} \times Y \hookrightarrow \mathbb{F} \times X \xrightarrow{\text{mult.}} X$$

is continuous, that is the full composite

$$\mathbb{F} \times Y \xrightarrow{\text{mult.}} Y \hookrightarrow X$$

is continuous. Hence multiplication: $\mathbb{F} \times Y \to Y$ is continuous by Proposition 1.21.
　As for quotients, a similar gimmick works.

$$\mathbb{F} \times (X/Y) \approx (\mathbb{F} \times X)/(0 \times Y) \qquad \text{[Thm. 1.23(i)]}$$

and

$$\mathbb{F} \times X \xrightarrow{\text{mult.}} X \xrightarrow{\text{proj.}} X/Y \text{ is continuous, that is}$$

$$\mathbb{F} \times X \xrightarrow{\text{proj.}} \quad (\mathbb{F} \times X)/(0 \times Y)$$

$$\wr\wr$$

$$\mathbb{F} \times (X/Y) \xrightarrow{\text{mult.}} X/Y$$

is continuous as a composite function, so multiplication: $\mathbb{F} \times (X/Y) \to X/Y$ is continuous by Theorem 1.23(c). $\square$

Proposition 2.3. *Suppose X is a topological vector space over $\mathbb{F}$, and $v_1, \ldots, v_n \in X$. Define $T : \mathbb{F}^n \to X$ by $T(c_1, \ldots, c_n) = \Sigma c_j v_j$. Then T is continuous.*

Proof. In fact,

$$\mathbb{F}^n \times X^n \approx \prod_{i=1}^{n}(\mathbb{F} \times X) \xrightarrow[\text{each factor}]{\text{mult. in}} \prod_{i=1}^{n} X \xrightarrow{\text{sum}} X$$

is jointly continuous on $\mathbb{F}^n \times X^n$, hence is separately continuous. Its continuity on $\mathbb{F}^n$ at $(v_1, \ldots, v_n) \in X^n$ gives our conclusion. $\square$

2.2 Special Subsets

When working with topological groups, some things become simpler when a neighborhood base at the identity consists of "symmetric" sets, that is sets for which $B = B^{-1}$. Here, that would read "$B = -B$," but it is not enough.

Definition 2.4. Suppose X is a vector space over $\mathbb{R}$ or $\mathbb{C}$. A subset $B \subset X$ is **balanced** if $\forall x \in B$, $\forall c$ in the base field with $|c| \leq 1$, we have $cx \in B$.

When the base field is $\mathbb{R}$, this just says that the line segment between x and $-x$ is in B, which implies our mental concept of "balanced." When the base field is $\mathbb{C}$, it says a good deal more, and many authors use the word "circled" in place of "balanced." Note that if $B \neq \emptyset$, then $0 \in B$ by taking $c = 0$.

Proposition 2.5. *Suppose X is a topological vector space.*

(a) If B is balanced, then so is B^-.
(b) The intersection of any family of balanced sets is balanced.
(c) If B is balanced, and $0 \in \text{int}(B)$, then $\text{int}(B)$ is balanced.
(d) Every neighborhood of 0 contains a balanced neighborhood of 0. In fact,
(e) X has a neighborhood base at 0 consisting of balanced sets, which can be taken to be all open, or all closed.

Proof. (a) If $0 \leq |c| \leq 1$, and $x \in B^-$, take a net $\langle x_\alpha \rangle$, $x_\alpha \in B$, with $x_\alpha \to x$. Multiplication by c is continuous, so $cx_\alpha \to cx$. Since $cx_\alpha \in B$, $cx \in B^-$.

(b) If B_α are balanced, and $x \in \cap B_\alpha$, and $|c| \leq 1$, then $\forall \alpha : x \in B_\alpha \Rightarrow cx \in B_\alpha$, so $cx \in B_\alpha$ for all α.

(c) If $0 < |c| \leq 1$, then multiplication by c is a homeomorphism which maps B into itself, so $c \cdot \text{int}(B)$ is an open subset of $cB \subset B$. Hence, if $x \in \text{int}(B)$, then $cx \in c \cdot \text{int}(B) \subset B$, so cx is interior to B. On the other hand, if $c = 0$, then $cx = 0 \in \text{int}(B)$ by assumption.

(d) Suppose V is a neighborhood of 0. Since scalar multiplication is jointly continuous, $\exists$ an open neighborhood W of 0 and a $\delta > 0$ such that $|c| < \delta$ and $x \in W \Rightarrow cx \in V$. Set

$$B = \bigcup_{0 < |c| < \delta} cW.$$

Note that $B \subset V$, B is open, and $0 \in B$ (since $0 \in$ every cW). If $x \in B$, and $0 < |c'| \le 1$, then $x \in cW$ with $|c| < \delta \Rightarrow c'x \in c'cW$, and $0 < |c'c| < \delta$. Hence $0 < |c'| \le 1 \Rightarrow c'x \in B$. Finally, if $x \in B$, then $0 \cdot x = 0 \in B$.

(e) Part (d) says that the balanced open sets form a neighborhood base at 0. Corollary 1.10 says that their closures also form a neighborhood base at 0, and it consists of balanced sets by part (a). $\square$

Note that a set may be balanced without having a balanced interior: Consider the bowtie in $\mathbb{R}^2$:

This section is concerned primarily with two kinds of special subsets. Balanced sets are one kind; the other kind are the bounded sets.

Definition 2.6. Suppose X is a topological vector space. A set B in X is **bounded** if, for every neighborhood V of 0, there is a scalar c such that $B \subset cV$.

Proposition 2.7. *Suppose X is a topological vector space.*

(a) If B is bounded, and V is a neighborhood of 0, then there is a real scalar $t_0 \ge 0$ such that $B \subset cV$ whenever $|c| \ge t_0$.

(b) If B is bounded, then so is B^-.

(c) If B and C are bounded, then so are $B \cup C$ and $B + C$.

(d) Compact sets are bounded.

(e) A set B is bounded if, and only if, whenever $x_n \in B$, and $c_n \to 0$ in the base field, necessarily $c_n x_n \to 0$ in X.

Proof. (a) Choose a balanced neighborhood W of 0 such that $W \subset V$. Then $B \subset c_0 W$ for some c_0. Set $t_0 = |c_0|$. If $|c| \ge t_0$, then $|c_0/c| \le 1$, so $(c_0/c)W \subset W$, that is $c_0 W \subset cW$. Hence $B \subset c_0 W \subset cW \subset cV$.

(b) If V is a neighborhood of 0, choose a closed neighborhood W of 0 with $W \subset V$, in accordance with Corollary 1.10. Choose c such that $B \subset cW$. Then $B^- \subset cW \subset cV$ since cW is closed.

(c) If $B \subset cV$ when $|c| \ge t_0$, and $C \subset cV$ when $|c| \ge t_1$, then $B \cup C \subset cV$ when $|c| \ge \max(t_0, t_1)$. As for $B + C$, given a neighborhood V of 0, choose neighborhoods W_1 and W_2 of 0 for which $W_1 + W_2 \subset V$ [condition (ii) in Proposition 1.8]. Choose t_0 such that $B \subset cW_1$ when $|c| \geq t_0$, and t_1 such that $C \subset cW_2$ when $|c| \geq t_1$. Set $t_2 = \max(t_0, t_1)$. Then $B \subset t_2W_1$, and $C \subset t_2W_2$, so $B + C \subset t_2W_1 + t_2W_2 = t_2(W_1 + W_2) \subset t_2V$.

(d) Suppose K is compact, and V is a neighborhood of 0. Choose a balanced open neighborhood W of 0 for which $W \subset V$. Then $\forall x \in K$, $\left(\frac{1}{n}\right)x \to 0$ since $\frac{1}{n} \to 0$ in the base field, so $\left(\frac{1}{n}\right)x \in W$ for large enough n. That is, $x \in nW$. So:

$$K \subset \bigcup_{n=1}^{\infty} nW.$$

Compactness says

$$K \subset \bigcup_{n=1}^{N} nW = NW, \text{ for some } N$$

since $n \leq N \Rightarrow (n/N)W \subset W \Rightarrow nW \subset NW$ since W is balanced. Hence $K \subset NW \subset NV$.

(e) First, suppose B is bounded, $x_n \in B$, and $c_n \to 0$ in the base field. Let V be any neighborhood of 0, and suppose $B \subset cV$ whenever $|c| \geq t_0$. $\exists N$ such that $n \geq N \Rightarrow |c_n| < \frac{1}{(t_0+1)}$. Suppose $n \geq N$. If $c_n = 0$, then $c_nx_n = 0 \in V$. If $c_n \neq 0$, then $|c_n^{-1}| > t_0 + 1 > t_0$, so $x_n \in B \subset c_n^{-1}V \Rightarrow c_nx_n \in V$. In all cases, $c_nx_n \in V$ when $n \geq N$.

Next, suppose B is not bounded. Then by part (a), $\exists$ a neighborhood V of 0 for which $B \not\subset$ any cV. In particular, $B \not\subset nV$. Choose $x_n \in B - nV$, and set $c_n = \frac{1}{n} \cdot x_n \notin nV \Rightarrow c_nx_n \notin V$, so $c_nx_n \not\to 0$. $\qquad\square$

Several things are worth noting. First of all, in a normed space, "bounded" means exactly what it always has: $B \subset n$ times the open unit ball when $\|x\| < n$ for all $x \in B$.

Next, the fact that boundedness can be characterized using sequences [part (e)] is a good deal more important than it looks. Also, it is more-or-less obvious that subsets of bounded sets are bounded, and that if B is bounded, then so is cB. (In fact, if B is balanced, then so is cB.) Finally, there is a subtlety buried in the proof of part (d) above: Singletons are compact, hence bounded, so if $n \in X$ and V is neighborhood of 0, then there is a t_0 for which $x \in cV$ whenever $|c| \geq t_0$. This property is called *absorbent*: A set B is **absorbent** if, for all $x \in X$, there exists a $t_0 \geq 0$ such that $x \in cB$ whenever $|c| \geq t_0$. This concept will be returned to later.

In a normed space, the open unit ball is open and bounded. This does not often happen. If B is bounded, and $x \in \text{int}B$, then $B + (-x)$ is a bounded neighborhood of 0 [Proposition 2.7(c)]. Consider:

Proposition 2.8. *Suppose B is a bounded neighborhood of 0 in a topological vector space X. Then $\{2^{-n}B : n = 1, 2, \ldots\}$ is a neighborhood base at 0. In particular, X is first countable.*

Proof. If V is a neighborhood of 0, then $\exists\, t_0 \geq 0$ such that $B \subset cV$ when $|c| \geq t_0$. Choose $n \in \mathbb{N}$ with $2^n \geq t_0$. Then $B \subset 2^n V$, so $2^{-n} B \subset V$.   □

In particular, "locally bounded" implies first countable. In fact, as we shall see later, a Hausdorff locally convex space is locally bounded if, and only if, its topology comes from a norm.

We close with an addendum to Proposition 2.3.

Proposition 2.9. *Suppose $\mathbb{F} = \mathbb{R}$ or $\mathbb{C}$. Then the product topology on $\mathbb{F}^n$ is the only Hausdorff topological vector space topology on $\mathbb{F}^n$.*

Proof. Let $\mathscr{T}_p$ denote the product topology, and $\mathscr{T}_0$ some other topology making $\mathbb{F}^n$ into a Hausdorff topological vector space. Then by Proposition 2.3, $(\mathbb{F}^n, \mathscr{T}_p) \to (\mathbb{F}^n, \mathscr{T}_0)$ is continuous, so $\mathscr{T}_0 \subset \mathscr{T}_p$. Hence every $\mathscr{T}_0$ neighborhood of 0 is a $\mathscr{T}_p$ neighborhood of 0. It suffices to show that a standard $\mathscr{T}_p$ neighborhood of 0, $\{v \in \mathbb{F}^n : \|v\| < r\}$, contains a $\mathscr{T}_0$ neighborhood of 0: Then, the $\mathscr{T}_0$ neighborhoods of 0 will form a base at 0 for $\mathscr{T}_p$ as well as $\mathscr{T}_0$; since the topology is determined by a base at 0 (Proposition 1.8), we will get that $\mathscr{T}_p = \mathscr{T}_0$.

Consider $K = \{v \in \mathbb{F}^n : \|v\| = r\}$. K is compact in the $\mathscr{T}_p$ topology, hence is compact in the $\mathscr{T}_0$ topology (since $\mathscr{T}_0 \subset \mathscr{T}_p$), hence is closed in the $\mathscr{T}_0$ topology (since it is Hausdorff). Since $0 \notin K$, $\exists$ a balanced $\mathscr{T}_0$ neighborhood V of 0 with $V \subset \mathbb{F}^n - K$. If $x \in V$ and $\|x\| > r$, then $(r/\|x\|)x \in V$ since V is balanced, giving $(r/\|x\|)v \in V \cap K$, a contradiction. Hence $x \in V \Rightarrow \|x\| < r$, that is $V \subset \{v \in \mathbb{F}^n : \|v\| < r\}$.   □

Corollary 2.10. *In any Hausdorff topological vector space, finite-dimensional subspaces are closed.*

Proof. It follows from Proposition 2.9 that there is exactly one way to topologize a finite-dimensional vector space over $\mathbb{R}$ or $\mathbb{C}$ and make it into a Hausdorff topological vector space, since any topology can be transported to $\mathbb{F}^n$ using a basis. (The map is in Proposition 2.3). This topology is the norm topology, which is first countable and sequentially complete, hence is complete (Theorem 1.34), hence is closed in any larger Hausdorff space (Proposition 1.30).   □

Corollary 2.11. *A locally compact Hausdorff topological vector space is finite-dimensional.*

Proof. Suppose X is a locally compact Hausdorff topological vector space. Let V be an open neighborhood of 0 for which V^- is compact (Proposition 1.6). Note that V^- is bounded, so V is bounded, so $\{2^{-n} V : n = 1, 2, \ldots\}$ is a neighborhood base at 0 (Proposition 2.8). Since V^- is compact,

$$V^- \subset \bigcup_{x \in X} x + \frac{1}{2} V \text{ implies}$$

$$V^- \subset \bigcup_{j=1}^{n} x_j + \frac{1}{2} V, \text{ some } x_1, \ldots, x_n \in X.$$

Let Y denote the span of $x_1, \ldots, x_n$; Y is finite-dimensional, hence is closed (Corollary 2.10). But $V \subset V^- \subset Y + \frac{1}{2}V$. Hence, by induction on n, $V \subset Y + 2^{-n}V$:

$$2^{-n}V \subset 2^{-n}Y + 2^{-(n+1)}V, \text{ so (induction hypothesis)}$$
$$V \subset Y + 2^{-n}V \subset Y + 2^{-n}Y + 2^{-(n+1)}V = Y + 2^{-(n+1)}V.$$

Hence

$$V \subset \bigcap_{n=1}^{\infty} Y + 2^{-n}V = Y^- = Y \text{ (Proposition 1.9)}.$$

But V is absorbent; if $x \in X$, then $x \in cV \subset cY \subset Y$ for some scalar c. Hence $X = Y$. $\qquad\qquad\square$

2.3 Convexity

Convex subsets are special subsets, but convexity rates a section all its own. There is a mythology that convexity only matters in functional analysis through local convexity. This is not true at all. Convexity plays a role in some of the oddest places.

If X is a vector space over $\mathbb{R}$ (or $\mathbb{C}$; any vector space over $\mathbb{C}$ is also a vector space over $\mathbb{R}$), and $C \subset X$, then C is convex when $\forall x, y \in C$, $\forall t \in [0, 1]$, $tx + (1 - t)y \in C$. Several things are clear. First, observe that when checking if $x, y \in C \Rightarrow tx(1-t)y \in C$ for $0 \leq t \leq 1$, we need only check for $0 < t < 1$, since x and y are in C by assumption. Also, if C is convex, then so is $x + aC$ whenever $x \in X$ and a is a scalar (even a complex scalar when X is a vector space over $\mathbb{C}$). Finally, note that the intersection of any family of convex sets is convex, so if $A \subset X$, then the intersection of all convex sets containing A will be the smallest convex set containing A; this set is called the *convex hull* of A, written as con(A). It has an internal construction as well, reminiscent of the subspace spanned by a set as either the intersection of all subspaces that contain it (external) or as the set of linear combinations (internal).

Proposition 2.12. *Suppose X is a vector space over $\mathbb{R}$, and $A \subset X$. Then the convex hull of A is the set of all "convex combinations," or sums:*

$$\sum_{i=1}^{n} t_i x_i : x_1, \ldots, x_n \in A; t_1, \ldots, t_n \in [0, 1], \sum_{i=1}^{n} t_i = 1.$$

In particular, if A is already convex, then all such sums lie in A.

Proof. The closing "In particular" is actually the starting point. Assume A is convex for a moment. Then all such sums must lie in A, by induction on n: If $n = 1$, then $t_1 = 1$ and $t_1 x_1 = x_1 \in A$. For the induction step, assume $t_1, \ldots, t_{n+1} \in [0, 1]$, $x_1, \ldots, x_{n+1} \in A$, and $\Sigma t_i = 1$. If $t_{n+1} = 1$, then all other $t_i = 0$, so $\Sigma t_i x_i = x_{n+1} \in A$. If $t_{n+1} \neq 1$, then

$$
\sum_{i=1}^{n+1} t_i x_i = (1 - t_{n+1}) \underbrace{\sum_{i=1}^{n} \frac{t_i}{1 - t_{n+1}} x_i}_{\text{in } A \text{ (induction hypothesis)}} + t_{n+1} x_n.
$$

Now to the general case. Since $\mathrm{con}(A)$ is convex, it must contain all convex combinations from $\mathrm{con}(A)$, so it will contain all convex combinations from A since $A \subset \mathrm{con}(A)$. It remains only to show that the set of all convex combinations is itself convex.

Suppose $x_1, \ldots, x_n \in A$; $y_1, \ldots, y_m \in A$; $t_i, \ldots, t_n \in [0, 1]$, and $t_1', \ldots, t_m' \in [0, 1]$, with $\Sigma t_i = \Sigma t_i' = 1$. If $0 \leq s \leq 1$, then

$$
s \sum_{i=1}^{n} t_i x_1 + (1 - s) \sum_{j=1}^{m} t_j' y_j' = \sum_{i=1}^{n} s t_i x_i + \sum_{j=1}^{m} (1 - s) t_j' y_j
$$

is itself a convex combination of length $n + m$, since

$$
\sum_{i=1}^{n} s t_i + \sum_{j=1}^{m} (1 - s) t_j' = s \sum_{i=1}^{n} t_i + (1 - s) \sum_{j=1}^{m} t_j' = s + (1 - s) = 1.
$$

$\square$

In general, convexity "plays well with others." It interacts well with topological considerations, even with nothing more than the basic topological vector space assumptions.

Proposition 2.13. *Suppose X is a topological vector space, and C is a convex subset of X. Then C^- and $\mathrm{int}(C)$ are also convex.*

Proof. C^- is convex: Suppose $x \in C$ and $y \in C^-$, and $0 \leq t \leq 1$. If $t = 0$ or 1, then $tx + (1 - t)y \in C^-$ trivially, so assume $0 < t < 1$. In accordance with Proposition 1.3(a), y is a limit of a net $\langle y_\alpha \rangle$ from C, and multiplication by $(1-t)$ is a homeomorphism, so $(1-t)y_\alpha \to (1-t)y$. Hence $tx+(1-t)y_\alpha \to tx+(1-t)y$ since $tx + \bullet$ is also a homeomorphism. But all $tx + (1-t)y_\alpha \in C$, so $tx + (1-t)y \in C^-$ again by Proposition 1.3(a).

Next, suppose $x \in C^-$ and $y \in C^-$, and $0 \leq t \leq 1$. If $t = 0$ or 1 then $tx + (1 - t)y \in C^-$ trivially, so assume $0 < t < 1$. In accordance with Proposition 1.3(a), x is a limit of a net $\langle x_\alpha \rangle$ from C, and $\cdots$. As above, $tx_\alpha + (1 - t)y \in C^-$ and $tx_\alpha + (1 - t)y \to tx + (1t - t)y$, so $tx + (1 - t)y \in C^-$.

Now for int(C). Suppose $x \in C$ and $y \in$ int(C). If $0 \le t < 1$, then

$$tx + (1-t)y \in tx + (1-t)\text{int}(C) \subset C;$$

so in particular, $tx + (1-t)y$ is an interior point of C. In particular, if $x \in$ int(C), then $tx + (1-t)y \in$ int(C) even when $t = 1$, since $t = 1 \Rightarrow tx + (1-t)y = x \in$ int(C). $\qquad\square$

The "$0 \le t < 1$" restriction in the proof is suggestive, and will play a rôle in later constructions. The generality covering much of that is the following:

Proposition 2.14. *Suppose X is a vector space over $\mathbb{R}$, and suppose C and D are two convex subsets of X, while I is a (possibly degenerate) subinterval of $[0, 1]$. Then*

$$E = \{tx + (1-t)y : x \in C, y \in D, t \in I\}$$

is convex.

Proof. Given $t, t' \in I$; $x, x' \in C$; and $y, y' \in D$; suppose $0 < s < 1$. If $st + (1-s)t' = 0$, then $0 = t = t' \in I$, and

$$s(tx + (1-t)y) + (1-s)(t'x' + (1-t')y')$$
$$= sy + (1-s)y' = 0 \cdot x + (1-0)(sy + (1-s)y') \in E.$$

Similarly, if $s(1-t) + (1-s)(1-t') = 0$, then $1 = t = t' \in I$, and

$$s(tx + (1-t)y) + (1-s)(t'x' + (1-t')y')$$
$$= sx + (1-s)x' = 1 \cdot (sx + (1-s)x') + 0 \cdot y \in E.$$

When $st + (1-s)t' > 0$ and $s(1-t) + (1-s)(1-t') > 0$:

$$s(tx + (1-t)y) + (1-s)(t'x' + (1-t')y')$$
$$= stx + (1-s)t'x' + s(1-t)y + (1-s)(1-t')y'$$
$$= (st + (1-s)t') \cdot \left(\frac{st}{st + (1-s)t'}x + \frac{(1-s)t'}{st + (1-s)t'}x' \right)$$
$$+ (s(1-t) + (1-s)(1-t')) \cdot$$
$$\left(\frac{s(1-t)}{s(1-t) + (1-s)(1-t')}y + \frac{(1-s)(1-t')}{s(1-t) + (1-s)(1-t')}y' \right)$$
$$= (st + (1-s)t')x'' + (1 - (st + (1-s)t'))y'' \in E$$

since $x'' = \dfrac{st}{st + (1-s)t'}x + \dfrac{(1-s)t'}{st + (1-s)t'}x' \in C$,

$$y'' = \frac{s(1-t)}{s(1-t) + (1-s)(1-t')}y + \frac{(1-s)(1-t')}{s(1-t) + (1-s)(1-t')}y' \in D,$$

$st + (1 - s)t' \in I$ since I is an interval, and

$st + (1 - s)t' + s(1 - t) + (1 - s)(1 - t') = s(t + 1 - t) + (1 - s)(t' + 1 - t')$
$= 1$, so that $1 - [st + (1 - s)t'] = s(1 - t) + (1 - s)(1 - t')$.

$\square$

The preceding is not pretty, but the "slick" proof (based on considerations from the next section) is just *too* slick, and looks like a scam (although it is not). See Exercise 3 for a description of this approach.

There is one more result of major importance, which concerns the interaction between "convex" and "balanced." These two concepts play particularly well together.

Theorem 2.15. *Suppose X is a topological vector space, and suppose B is a convex, balanced subset of X. Then the following are equivalent:*

(i) int$(B) \neq \emptyset$.
(ii) $0 \in$ int(B).
(iii) int$(B) = [0, 1)B$ and $B \neq \emptyset$.

Proof. Since $B \neq \emptyset \Rightarrow 0 \in 0 \cdot B \subset B$: (iii) $\Rightarrow$ (ii); while (ii) $\Rightarrow$ (i) trivially. To show (i) $\Rightarrow$ (iii), assume $x \in B$, $y \in$ int(B), and $0 \leq t < 1$. Then $-y \in B$ since B is balanced, so by Proposition 2.12:

$$tx = tx + (1 - t)\left(\frac{1}{2}y + \frac{1}{2}(-y)\right)$$

$$\in tx + \frac{1-t}{2}(-y) + \frac{(1-t)}{2}\text{int}(B) \subset B$$

and $\frac{1-t}{2}$int(B) is open since $\frac{1-t}{2} > 0$. $\square$

It is difficult to overemphasize the usefulness of this result: Solely given that int(B) was nonempty, we have a formula for its interior which makes *no* reference to the topology. Condition (ii) will play its own role later.

We close this section with a look back at the L^p-spaces, $0 < p < 1$. The failure of local convexity here is usually developed using the Hahn–Banach theorem, but it is, in fact, much more fundamental.

Proposition 2.16. *Suppose μ is Lebesgue measure on $[0, 1]$, and $0 < p < 1$. Suppose C is a convex subset of $L^p(\mu)$. Then int$(C) \neq \emptyset \Rightarrow C = L^p(\mu)$.*

Proof. If $[h] \in$ int(C), then $[0] \in$ int$(C) - [h] =$ int$(C - [h])$, the latter equality holding since translation is a homeomorphism. Choose $r > 0$ so that $\rho([f]) < r \Rightarrow [f] \in$ int$(C - [h])$, where as before,

$$\rho([f]) = \int_0^1 |f(t)|^p d\mu(t).$$

Suppose $[f] \in L^p(\mu)$. Set

$$F(x) = \int_0^x |f(t)|^p d\mu(t).$$

Then F is a nondecreasing continuous function on $[0, 1]$, with $F(1) = \rho([f])$. Choose an integer N such that $N^{1-p} > \rho([f])/r$, and $0 = x_0 < x_1 < \cdots \le x_N = 1$ so that $F(x_k) = \rho([f])k/N$ (intermediate value theorem).

Set $I_k = (x_{k-1}, x_k]$ for $1 < k \le N$, and $I_1 = [x_0, x_1]$. Note that $\Sigma \chi_{I_k} = 1$, where χ_{I_k} is the characteristic function of I_k. But now

$$[f] = \sum_{k=1}^N \frac{1}{N} \cdot (N[f \chi_{I_k}]), \text{ and}$$

$$\rho(N[f \chi_{I_k}]) = N^p \int_{x_{k-1}}^{x_k} |f(t)|^p dt$$

$$= N^p (F(x_k) - F(x_{n-1}))$$

$$= N^p \rho([f])/N = \rho([f])/N^{1-p} < r.$$

Hence $N[f\chi] \in C - [h]$ by the choice of r, so $[f] \in C - [h]$ by Proposition 2.12. Since $[f]$ was arbitrary, $C - [h] = L^p(\mu)$, so $C = L^p(\mu)$. □

2.4 Linear Transformations

This section will primarily wrap up some odds and ends concerning the effects of linear transformations, and will close with a technical result to be used later.

When it comes to balanced, convex, or absorbent sets, the topology plays no rôle.

Proposition 2.17. *Suppose X and Y are vector spaces over $\mathbb{R}$ or $\mathbb{C}$, and suppose T is a linear transformation from X to Y.*

(a) If B is convex in X, then $T(B)$ is convex in Y.
(b) If B is balanced in X, then $T(B)$ is balanced in Y.
(c) If C is convex in Y, then $T^{-1}(C)$ is convex in X.
(d) If C is balanced in Y, then $T^{-1}(C)$ is balanced in X.
(e) If C is absorbent in Y, then $T^{-1}(C)$ is absorbent in X.

Proof. The underlying ideas are pretty simple:

(a) If $T(x), T(y) \in T(B)$; $x, y \in B$; and $0 \le t \le 1$; then $tT(x) + (1-t)T(y) = T(tx + (1-t)y) \in T(B)$.

(b) If $T(x) \in T(B)$, $x \in B$, and $|c| \le 1$, then $cT(x) = T(cx) \in T(B)$.

(c) If $x, y \in T^{-1}(C)$, and $0 \le t \le 1$, then $T(tx + (1-t)y) = tT(x) + (1-t)T(y) \in C$, so $tx + (1-t)y \in T^{-1}(C)$.

(d) If $x \in T^{-1}(C)$, and $|c| \leq 1$, then $T(cx) = cT(x) \in C$, so $cx \in T^{-1}(C)$.
(e) If $x \in X$, and $T(x) \in cC$ whenever $|c| \geq t_0$, then $x \in T^{-1}(cC) = cT^{-1}(C)$
whenever $|c| \geq t_0$.

$\square$

That was quick. The next result is even quicker.

Proposition 2.18. *Suppose X and Y are topological vector spaces, and T is a continuous linear transformation from X to Y. If B is a bounded subset of X, then $T(B)$ is a bounded subset of Y.*

Proof. If V is a neighborhood of 0 in Y, then $T^{-1}(V)$ is a neighborhood of 0 in X, so $B \subset cT^{-1}(V) = T^{-1}(cV)$ for some c, whence $T(B) \subset cV$. $\square$

A linear transformation T from a topological vector space X to another topological vector space Y will be called *bounded* if $T(B)$ is bounded in Y whenever B is bounded in X. Proposition 2.18 just says that continuous linear transformations are bounded. The converse is not true in general, but the issue can and will be addressed for locally convex spaces in Chap. 4.

The technical result really does not belong here, since there are no linear transformations used—or are there? Actually, it is based on the fact that complex scalar multiplication is a real-linear transformation. The technical condition at the end is a preview of the verb "to absorb": A set B *absorbs* a set A if there is a real scalar t_0 such that $A \subset cB$ whenever $|c| \geq t_0$. This accords with earlier terminology: An absorbent set is one that absorbs points. We will eventually need much more latitude concerning which sets absorb what.

Proposition 2.19. *Suppose X is a vector space over $\mathbb{C}$, and suppose B is a nonempty convex subset of X that is $\mathbb{R}$-balanced, that is $tB \subset B$ for $-1 \leq t \leq 1$, and suppose $\mathscr{F}$ is a family of subsets of X such that for all $A \subset \mathscr{F}$: $cA \in \mathscr{F}$ for all $c \in \mathbb{C}$, and $A \subset tB$ for some $t \in \mathbb{R}$. Then*

$$C = \bigcap_{0 \leq \theta < 2\pi} e^{i\theta} B$$

is convex and $\mathbb{C}$-balanced, and $\forall\, A \in \mathscr{F}\, \exists\, t_0 \in \mathbb{R}$ with $A \subset cC$ when $|c| \geq t_0$.

Proof. Note that $\{0\} = 0 \cdot B \subset B$, so $0 \in C$. It is also immediate that C is convex (since it is an intersection of convex sets).

C is balanced: If $x \in C$, and $|c| \leq 1$, write $c = re^{i\phi}$, $r \geq 0$. Then for $0 \leq \theta < 2\pi$,

$$x \in e^{i(\theta-\phi)} B \Rightarrow e^{i(\phi-\theta)} x \in B$$
$$\Rightarrow re^{i(\phi-\theta)} x \in B \qquad (B \text{ is } \mathbb{R}\text{-balanced})$$
$$\Rightarrow ce^{-i\theta} x \in B \Rightarrow cx \in e^{i\theta} B.$$

The real question is why C absorbs members of $\mathscr{F}$. The underlying trick involves the convexity. First, note that the convex hull of 2, $-1 - i\sqrt{3}$, and $-1 + i\sqrt{3}$ in $\mathbb{C}$ contains the unit disk:

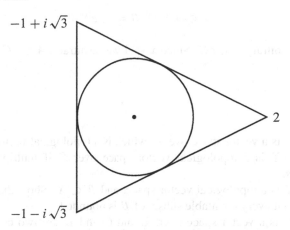

$$-1 + i\sqrt{3}$$

$$2$$

$$-1 - i\sqrt{3}$$

If $A \in \mathscr{F}$, $A \subset \{0\}$, then $A \subset cC$ for all $c \in \mathbb{C}$, so suppose $A \in \mathscr{F}$, $A \not\subset \{0\}$. Choose t_1, t_2, and t_3, so that $2A \subset t_1 B$, $(-1 + i\sqrt{3})A \subset t_2 B$, and $(-1 - i\sqrt{3})A \subset t_3 B$. Note that t_1, t_2, and t_3 are nonzero since $0 \cdot B = \{0\}$, and we are assuming that $A \not\subset \{0\}$. Set $t_0 = \max(|t_1|, |t_2|, |t_3|) > 0$, and suppose $|c| \geq t_0$. Write $c = re^{i\theta}$; then $r \geq t_0 > 0$. Fix any ϕ, $0 \leq \phi < 2\pi$, and choose $s_1, s_2, s_3 \geq 0$, $s_1 + s_2 + s_3 = 1$, so that

$$e^{-i(\theta + \phi)} = s_1 \cdot 2 + s_2 \cdot (-1 + i\sqrt{3}) + s_3 \cdot (-1 - i\sqrt{3}).$$

Note that $|t_i/t_0| \leq 1$ for $i = 1, 2, 3$; and $0 < t_0/r \leq 1$; so $|t_i/r| \leq 1$ for $i = 1, 2, 3$. Thus, since B is $\mathbb{R}$-balanced, for any $x \in A$:

$$2x \in t_1 B \Rightarrow \frac{2}{t_1} x \in B \Rightarrow \frac{t_1}{r} \cdot \frac{2}{t_1} x \in B, \text{ that is } \frac{2}{r} x \in B;$$

$$(-1 + i\sqrt{3})x \in t_2 B \Rightarrow \frac{-1 + i\sqrt{3}}{t_2} x \in B \Rightarrow \frac{t_2}{r} \frac{-1 + i\sqrt{3}}{t_2} x \in B,$$

that is $\frac{-1 + i\sqrt{3}}{r} x \in B$; and

$$(-1 - i\sqrt{3})x \in t_3 B \Rightarrow \frac{-1 - i\sqrt{3}}{t_3} x \in B \Rightarrow \frac{t_3}{r} \frac{-1 - i\sqrt{3}}{t_3} x \in B,$$

that is $\frac{-1 - i\sqrt{3}}{r} x \in B$. Thus, since B is convex,

$$s_1 \frac{2}{r} x + s_2 \frac{-1 + i\sqrt{3}}{r} x + s_3 \frac{-1 - i\sqrt{3}}{r} x \in B, \text{ that is}$$

$$r^{-1}(s_1 \cdot 2 + s_2 \cdot (-1 + i\sqrt{3}) + s_3 \cdot (-1 - i\sqrt{3}))x \in B, \text{ that is}$$

$$r^{-1}e^{-i(\theta+\phi)}x \in B, \text{ that is}$$

$$x \in re^{i(\theta+\phi)}B = c \cdot e^{i\phi}B.$$

Since ϕ was arbitrary, $x \in cC$. Since $x \in A$ was arbitrary, $A \subset cC$. $\square$

Exercises

1. Suppose X is a vector space over $\mathbb{C}$ which is a topological vector space over $\mathbb{R}$. Show that X is a topological vector space over $\mathbb{C}$ if multiplication by i is continuous.
2. Suppose X is a topological vector space, and $B \subset X$. Show that B is bounded provided that every countable subset of B is bounded.
3. Suppose X is a vector space over $\mathbb{R}$, and C and D are two convex subsets of X, while I is a subinterval of $[0, 1]$. Suppose $x, x' \in C$ and $y, y' \in D$. Let $L_1 = \{tx + (1-t)x' : 0 \le t \le 1\}$ and $L_2 = \{ty + (1-t)y' : 0 \le t \le 1\}$. Show that

$$E_0 = \{tx'' + (1-t)y'' : x'' \in L_1, y'' \in L_2, t \in I\}$$

is equal to $x + T(E_1)$, where $T : \mathbb{R}^3 \to X$ is a linear transformation defined by

$$T((1,0,0)) = x' - x$$

$$T((0,1,0)) = y - x$$

$$T((0,1,1)) = y' - x$$

and E_1 is the (obviously convex) intersection of the tetrahedron with corners $(0,0,0)$, $(1,0,0)$, $(0,1,0)$, and $(0,1,1)$ with the slab $\{(x, y, z) \in \mathbb{R}^3 : y \in I\}$:

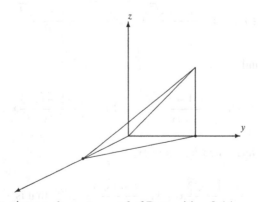

Use this to give an alternate proof of Proposition 2.14.

4. Suppose B is a convex subset of a topological vector space X, and suppose $0 \in$ B, $\mathrm{int}(B) \neq \emptyset$, and $\forall x \in X \exists \epsilon > 0$ with $-\epsilon x \in B$. Show that $\mathrm{int}(B) = [0, 1)B$. *Hint:* Mimic the proof of Theorem 2.15. You do not have "$-y \in B$," but you do have some $\epsilon \cdot (-y) \in B$.

5. Show that $L^p(\mu)$ is complete when $0 < p < 1$. (Use Theorem 1.35 and Exercise 16 from Chap. 1. Note that

$$\left(\sum_{k=1}^{n} |f_k(t)| \right)^p \leq \sum_{k=1}^{n} |f_k(t)|^p \nearrow \sum_{k=1}^{\infty} |f_k(t)|^p.$$

6. Suppose Y is a closed subspace of $L^p(\mu)$, where $\mu = $ Lebesgue measure on $[0, 1]$ and $0 < p < 1$, and suppose $L^p(\mu)/Y$ is finite-dimensional. Show that $Y = L^p(\mu)$. *Hint:* If not, apply Proposition 2.9 to produce an open convex subset violating Proposition 2.16.

7. Suppose X is a topological vector space, and Y is a subspace. Show that Y^- is a subspace. (This is almost too easy: You already know that Y^- is an additive subgroup [Proposition 1.24], while multiplication by nonzero scalars are homeomorphisms.)

Chapter 3
Locally Convex Spaces

3.1 Bases

Once again, a topological vector space will be called a locally convex space if it is locally convex, that is, if each point has a neighborhood base consisting of convex sets. The Hausdorff condition will also need to be imposed to make use of this (in fact, Schaefer [33] imposes that condition on locally convex spaces without imposing it on topological vector spaces), but there is some value in allowing intermediate considerations of locally convex spaces that are not Hausdorff.

The reader is assumed to be familiar with Banach spaces. Here is an example manufactured from standard Banach spaces which is not itself a normed space.

Example 4. Let m denote the Lebesgue measure on $[0, 1]$. Recall that if $p < q$, then $L^p(m) \supset L^q(m)$ in this case, with $\|f\|_p \leq \|f\|_q$ for $f \in L^q(m)$: writing $1 = 1/(q/p) + 1/r$, and noting that $|f|^p \in L^{q/p}(m)$, Hölder's inequality says that

$$\|f\|_p^p = \int_0^1 |f|^p \cdot 1 \, dm \leq \| \, |f|^p \|_{q/p} \cdot \|1\|_r$$

$$= \left(\int_0^1 (|f|^p)^{q/p} dm \right)^{p/q} = \|f\|_q^p.$$

Set

$$X = \bigcap_{p \geq 1} L^p(m), \text{ and}$$

$$\mathscr{B}_0 = \{\{[f] \in X : \|f\|_p \leq r\} : 0 < r < \infty, 1 \leq p < \infty\}.$$

$\mathscr{B}_0$ is a base at 0 for a topology on X that is not a norm topology but is manufactured using norms. By the way, while $L^\infty(m) \subset X$, these spaces are not the same, since $[\ln(x)] \in X$, but $\ln(x)$ is not essentially bounded.

M.S. Osborne, *Locally Convex Spaces*, Graduate Texts in Mathematics 269, DOI 10.1007/978-3-319-02045-7_3, © Springer International Publishing Switzerland 2014

The preceding example will reappear in Sect. 3.7. In fact, there is nothing wrong about looking ahead at Sects. 3.7 and 3.8 (after a peek at Theorem 3.2 below). After all, this is not a mystery novel.

A neighborhood base at 0 can be "translated" to any point, so, as usual, we concentrate on a base at 0. We will want the "balanced" condition, so our first result is aimed at guaranteeing that we are not working in a vacuum.

Proposition 3.1. *Suppose X is a locally convex space. Then X has a neighborhood base at* 0 *consisting of convex, balanced sets that can all be taken to be open or all closed.*

Proof. Suppose U is open and $0 \in U$. There is a convex set C with $0 \in \text{int}(C)$ and $C \subset U$, since X is locally convex. There is a balanced, open set W with $0 \in W \subset \text{int}(C)$ by Proposition 2.5(e). Let V denote the convex hull of W. Then $V \subset \text{int}(C)$, since $\text{int}(C)$ is convex (Proposition 2.13). V is also convex (it is a convex hull), and $0 \in \text{int}(V)$, since $0 \in W \subset V$ and W is open. It remains to show that V is balanced; it will then follow that $\text{int}(V)$ is convex (Proposition 2.13), balanced [Proposition 2.5(c)], and $0 \in \text{int}(V) \subset C \subset U$. From this, it will follow that (by letting U float) the set of all open, convex, balanced sets containing 0 forms a neighborhood base at 0. To get a neighborhood base at 0 consisting of closed, convex, balanced sets, just take closures [Corollary 1.10, plus Propositions 2.13 and 2.5(a)].

V is balanced: suppose $y \in V$ and $|c| \leq 1$. Choose $x_1, \ldots, x_n \in W$ with $y = \Sigma t_i x_i$, $t_i \geq 0$, $\Sigma t_i = 1$. Then $cy = \Sigma t_i c x_i \in \text{con}(W) = V$, since each $c x_i \in W$ (W is balanced). $\qquad\Box$

Next, we want to specify when we have a neighborhood base at 0. Actually, while an exhaustive set of conditions can be given, the following will suffice for every (yes, *every*) construction of a locally convex space in this book that is not already covered (e.g., subspaces, quotients, and products). For terminology, recall that a subset B of a vector space X over $\mathbb{R}$ or $\mathbb{C}$ is *absorbent* if, for all $x \in X$, there exists $t_0 > 0$ such that $x \in cB$ whenever $|c| \geq t_0$.

Theorem 3.2. *Suppose X is a vector space over $\mathbb{R}$ or $\mathbb{C}$, and suppose $\mathscr{B}_0$ is a nonempty family of convex, balanced, absorbent sets satisfying the following two conditions:*

(α) *If $B \in \mathscr{B}_0$, then $\frac{1}{2}B \in \mathscr{B}_0$.*
(β) *If $B_1, B_2 \in \mathscr{B}_0$, then $\exists B_3 \subset \mathscr{B}_0$ with $B_3 \subset B_1 \cap B_2$.*

Then $\mathscr{B}_0$ is a base at 0 *for a (unique) topology on X, making X into a locally convex space. This space is Hausdorff if, and only if, $\bigcap \mathscr{B}_0 = \{0\}$.*

Proof. First note that we do get a topological group via Proposition 1.8: running through conditions (i)–(iv) there:

(i)✓ $-B = B$ when $B \in \mathscr{B}_0$ (balanced condition);
(ii)✓ $\frac{1}{2}B + \frac{1}{2}B \subset B$ when $B \in \mathscr{B}_0$ (B is convex);
(iii)✓ $(X, +)$ is commutative;
(iv)✓ This is condition (β).

Also, via Corollary 1.11, the last sentence in this theorem is validated, so it only remains to show that scalar multiplication is jointly continuous.

Let $\mathbb{F}$ denote the base field, and suppose $x_0 \in X$, $c_0 \in \mathbb{F}$, and $c_0 x_0 = y_0 \in U$, with U open. Choose $B \subset \mathcal{B}_0$ with $y_0 + B \subset U$, and choose $n \in \mathbb{N}$ so that $2^n > |c_0|$. Now $2^{-k} B \in \mathcal{B}_0$ for $k = 1, 2, \ldots$, by an easy induction on k. Also, there is a $t_0 > 0$ such that $x_0 \in cB$ when $|c| \geq t_0$ since B is absorbent. Choose $m \in \mathbb{N}$ so that $2^m \geq t_0$. Then $x_0 \in 2^m B$, so $2^{-m} x_0 \in B$.

Choose $\epsilon > 0$ so that $|c - c_0| < \epsilon \Rightarrow |c| < 2^n$, and $\epsilon \leq 2^{-m-1}$. Suppose $|c - c_0| < \epsilon$, and $x \in x_0 + 2^{-n-1} B$. Then $x - x_0 \in 2^{-n-1} B$, so:

$$c(x - x_0) \in c2^{-n-1} B = \frac{c}{2^n} \cdot \frac{1}{2} B \subset \frac{1}{2} B$$

since $\left|\frac{c}{2^n}\right| < 1$ and $\frac{1}{2} B$ is balanced; and

$$(c - c_0)x_0 = \left(2^{m+1}(c - c_0)\right) \cdot \frac{1}{2} \cdot 2^{-m} x_0$$

$$\in \left(2^{m+1}(c - c_0)\right) \cdot \frac{1}{2} B \subset \frac{1}{2} B$$

since $\left|2^{m+1}(c - c_0)\right| < 1$ and $\frac{1}{2} B$ is balanced.
Hence:

$$cx - y_0 = cx - c_0 x_0 = cx - c x_0 + c x_0 - c_0 x_0$$

$$= c(x - x_0) + (c - c_0)x_0 \in \frac{1}{2} B + \frac{1}{2} B \subset B$$

since B is convex.

Thus $cx \in y_0 + B \subset U$, so

$$\{c : |c - c_0| < \epsilon\} \times \left(x_0 + \text{int}(2^{-n-1} B)\right)$$

is a neighborhood of (c_0, x_0) in $\mathbb{F} \times X$ that scalar multiplication maps into U. □

3.2 Gauges and Seminorms

It is clear that convexity is primarily a geometric condition. To make analytic use of convexity, this geometry must somehow be translated into some kind of analytic, that is, functional, property. The context for this does not start with just any convex set, but it does make strong use of convexity itself.

Start with a vector space X over $\mathbb{R}$ and a convex set C, with $0 \in C$. If $x \in X$, define $\varphi_x : \mathbb{R} \to X$ by $\varphi_x(t) = tx$; then $\varphi_x^{-1}(C)$ is a convex subset of $\mathbb{R}$ [Proposition 2.17(c)]. That is, $\varphi_x^{-1}(C)$ is an interval, and $0 \in \varphi_x^{-1}(C)$ since $0 \in C$. The assumption we make is that, for all $x \in X$, the interval $\varphi_x^{-1}(C)$ has 0 as an interior point, that is, some $(-\epsilon, \epsilon) \subset \varphi_x^{-1}(C)$. In particular, $(\epsilon/2)x \in C$, so $x \in (2/\epsilon)C$. This assumption is sometimes referred to by saying that 0 is an "internal point" of C, a terminology we use here, for a while at least. Note that 0 is an internal point of C if and only if C is absorbent (Proposition 2.19 applied to $C \cap (-C)$ and $\mathscr{F} = $ all singletons, when the base field is $\mathbb{C}$). Also, note that if X is a topological vector space, and $0 \in \text{int}(C)$, then 0 is an internal point since $\varphi_x^{-1}(\text{int}(C))$ actually *is* an open interval.

The geometric/analytic crossover is incorporated in the following proposition.

Proposition 3.3. *Suppose X is a vector space over $\mathbb{R}$, and C is a convex subset of X with $0 \in C$. Then:*

(a) *If $0 < s < t$, then $sC \subset tC$.*
(b) *If $s > 0$ and $t > 0$, then $sC + tC = (s+t)C$.*
(c) *If 0 is actually an internal point of C, then $I_x = \{t > 0 : x \in tC\}$ is a semi-infinite interval of the form $[a, \infty]$ or (a, ∞), for all $x \in X$.*

Proof. (a) If $x \in C$, then

$$sx = t \cdot \left(\frac{s}{t}x + \left(1 - \frac{s}{t}\right) \cdot 0 \right) \in tC.$$

(b) If $x \in C$, then $(s+t)x = sx + tx \in sC + tC$, so $(s+t)C \subset sC + tC$. If $x \in C$ and $y \in C$, then

$$sx + ty = (s+t)\left(\frac{s}{s+t}x + \frac{t}{s+t}y \right) \in (s+t)C.$$

(c) Since 0 is an internal point of C, I_x is nonempty by the earlier discussion that put $2/\epsilon$ in I_x. But now I_x becomes a Dedekind cut by part (a), since $t > s \in I_x \Rightarrow x \in sC \subset tC \Rightarrow t \in I_x$.   □

Corollary 3.4. *Suppose X is a vector space over $\mathbb{R}$, and C is a convex subset of X with $0 \in C$. Suppose 0 is an internal point of C. Then, in the notation of Proposition 3.3:*

(a) *If $x \in X$, and $s > 0$, then $I_{sx} = sI_x$.*
(b) *If $x, y \in X$, then $I_{x+y} \supset I_x + I_y$.*

Proof. (a) $t \in I_x \Leftrightarrow x \in tC \Leftrightarrow sx \in stC \Leftrightarrow st \in I_{sx}$.
(b) If $s \in I_x$ and $t \in I_y$, then $x \in sC$ and $y \in tC$, so $x+y \in sC+tC = (s+t)C$, so $s+t \in I_{x+y}$.   □

Now suppose, as above, that X is a real vector space, and C is a convex subset of X with $0 \in C$; suppose 0 is an internal point of C. The left endpoint of the interval I_x [the a in Proposition 3.3(c)] will be denoted by $p_C(x)$. Corollary 3.4(a) says that $p_C(sx) = sp_C(x)$ when $s > 0$. Also, $p_C(0) = 0$, since $I_0 = (0, \infty)$, so $p_C(sx) = sp_C(x)$ for $s \geq 0$. Also, Corollary 3.4(b) says that $p_C(x + y) \leq p_C(x) + p_C(y)$, since $I_x + I_y$ is a semi-infinite interval with the left endpoint $p_C(x) + p_C(y)$. That is, p_C is a *gauge*.

Definition 3.5. Suppose X is a vector space over $\mathbb{R}$ or $\mathbb{C}$. A **gauge** is a function $p : X \to \mathbb{R}$ satisfying the following two conditions:

(i) $p(tx) = tp(x)$ for all $t \geq 0$ and all $x \in X$.
(ii) $p(x + y) \leq p(x) + p(y)$ for all $x, y \in X$.

A gauge p is called a **seminorm** if p is nonnegative and $p(cx) = |c|p(x)$ for all $x \in X$ and c in the base field. A seminorm p is a **norm** if $p(x) = 0 \Rightarrow x = 0$.

Note that a gauge need *not* be nonnegative, although our functions p_C are. See Exercise 1 for a nice example.

Going back to our convex set C having 0 as an internal point, the function p_C has three names in the literature: the Minkowski gauge, the Minkowski functional, and the support function of C. We will use "Minkowski functional."

Since our space will be topological, one point needs to be made immediately.

Proposition 3.6. *Suppose X is a topological vector space over $\mathbb{R}$, and $p : X \to \mathbb{R}$ is a gauge. The following are equivalent:*

(i) p is continuous.
(ii) p is continuous at 0.
(iii) 0 is interior to $\{x \in X : p(x) \leq 1\}$.

Proof. (i) $\Rightarrow$ (ii) is trivial, and (ii) $\Rightarrow$ (iii) is as well, since the real number 0 is interior to $(-\infty, 1]$.

(iii) $\Rightarrow$ (i): Given any $x_0, x \in X$:

$$p(x) = p(x - x_0 + x_0) \leq p(x - x_0) + p(x_0), \text{ and}$$

$$p(x_0) = p(x_0 - x + x) \leq p(x_0 - x) + p(x), \text{ so}$$

$$- p(x_0 - x) \leq p(x) - p(x_0) \leq p(x - x_0).$$

Set $V = \text{int}\{y \in X : p(y) \leq 1\}$. If $x \in x_0 + \epsilon(V \cap (-V))$, that is, $x - x_0 \in \epsilon(V \cap (-V))$, then $x - x_0 \in \epsilon V$, that is, $x - x_0 = \epsilon y$, with $y \in V$, so $p(x - x_0) = p(\epsilon y) = \epsilon p(y) \leq \epsilon$, while $x - x_0 \in \epsilon(-V)$, that is, $x_0 - x \in \epsilon V$, that is, $x_0 - x = \epsilon z$ with $z \in V$, so $p(x_0 - x) = p(\epsilon z) = \epsilon p(z) \leq \epsilon$. That is, for $x \in x_0 + \epsilon(V \cap (-V))$:

$$-\epsilon \leq -p(x_0 - x) \leq p(x) - p(x_0) \leq p(x_0 - x) \leq \epsilon, \text{ or } |p(x) - p(x_0)| \leq \epsilon.$$

Since $\epsilon(V \cap (-V))$ is a neighborhood of 0, we are done. $\qquad \square$

Why ≤ 1? The following settles that and puts the whole business together.

Theorem 3.7. *Suppose X is a vector space over $\mathbb{R}$ or $\mathbb{C}$, and C is a convex subset of X with $0 \in C$. Suppose 0 is an internal point of C. Then the Minkowski functional p_C is a gauge, and*

$$\{x \in X : p_C(x) < 1\} \subset C \subset \{x \in X : p_C(x) \le 1\}.$$

Also, p_C is a seminorm if C is balanced. Finally, if X is a topological vector space, then $0 \in \text{int}(C)$ if, and only if, p_C is continuous.

Proof. The fact that p_C is a gauge was established earlier. As for the containments:

$$p_C(x) < 1 \Rightarrow 1 \in I_x \Rightarrow p_C(x) \le 1$$

$$\updownarrow$$

$$x \in C$$

Continuity of $p_C \Rightarrow \{x \in X : p_C(x) < 1\}$ is open, making 0 interior to C since $\{x \in X : p_C(x) < 1\} \subset C$. If $0 \in \text{int}(C)$, then 0 is interior to $\{x \in X : p_C(x) \le 1\}$ since $C \subset \{x \in X : p_C(x) \le 1\}$: this forces p_C to be continuous by Proposition 3.6.

There remains the seminorm condition when C is balanced, so assume C is balanced. If $|s| = 1$, then $sC \subset C$, and $s^{-1}C \subset C$, so $C = s^{-1}(sC) \subset s^{-1}C$. That is, $C = s^{-1}C$. But if $x \in X$, then $x \in tC \Leftrightarrow x \in ts^{-1}C \Leftrightarrow sx \in tC$. That is, $I_x = I_{sx}$. Hence $p_C(sx) = p_C(x)$ when $|s| = 1$. If c is in the base field, choose s with $|s| = 1$ so that $c = s|c|$. Then $p_C(cx) = p_C(s|c|x) = p_C(|c|x) = |c|p_C(x)$ since p_C *is* a gauge. $\qquad\qquad\square$

Note: $C = [-1, 1)$ is not balanced in $\mathbb{R}$, but its Minkowski gauge is $p_C(x) = |x|$, which is a (semi)norm.

Locally convex spaces are often defined using seminorms. In fact, this can be efficiently done using Theorem 3.2. Let X be a vector space over $\mathbb{R}$ or $\mathbb{C}$, and suppose $\mathscr{F}$ is a family of seminorms on X. One can take the following sets:

$$B = \bigcap_{j=1}^{n}\{x \in X : p_j(x) < 2^{-m}\}$$

with $n \in \mathbb{N}$ and $p_1, \ldots, p_n \in \mathscr{F}$; the collection of all such B's forms a base for a locally convex topology. In fact, Reed and Simon [29] actually define the term "locally convex space" in this way. This is not overly restrictive, since one can start with a base for the topology at 0 consisting of convex, balanced sets; take their Minkowski functionals; and apply this construction. However, it is somewhat more natural in many situations *not* to use seminorms. We shall return to this in Sect. 3.7, where the seminorm $\rightarrow$ locally convex space construction will only arise directly when the family of seminorms is countable. This situation *does* happen frequently.

Our next topic is the Hahn–Banach theorem, which deserves its own section. It concerns the extension of a linear functional bounded by a gauge. (Recall that a linear functional on a vector space is a linear transformation from the space to its base field.) When that gauge is a Minkowski functional, the following result is quite useful.

Proposition 3.8. *Suppose X is a vector space over $\mathbb{R}$, and C is a convex subset of X with $0 \in C$. Suppose 0 is an internal point of C, and p_C is the associated Minkowski functional. Finally, suppose $f : X \to \mathbb{R}$ is a linear functional. Then $f(x) \le p_C(x)$ for all $x \in X$ if, and only if, $f(y) \le 1$ for all $y \in C$.*

Proof. In the notation of Proposition 3.3, $p_C(x)$ is the left endpoint of the semi-infinite interval I_x, so $p_C(x)$ is the greatest lower bound for I_x. This is exactly what we need.

First, suppose $f(x) \le p_C(x)$ for all $x \in X$. If $y \in C$, then $y \in 1 \cdot C$, so $1 \in I_y$ and $p_C(y) \le 1$ since $p_C(y)$ is a lower bound for I_y. Thus $1 \ge p_C(y) \ge f(y)$.

Next, suppose $f(y) \le 1$ whenever $y \in C$. Fix $x \in X$, and suppose $t \in I_x$. Then $t > 0$ by definition of I_x. Also, $x \in tC$, so that $t^{-1}x \in C$ and (setting $y = t^{-1}x$) $f(t^{-1}x) \le 1$. But now $1 \ge f(t^{-1}x) = t^{-1}f(x)$, so $t \ge f(x)$. This all shows that $f(x)$ is a lower bound for I_x, so $f(x) \le p_C(x)$ since $p_C(x)$ is the greatest lower bound for I_x.   □

3.3 The Hahn–Banach Theorem

The Hahn–Banach theorem is fundamental to functional analysis. In a sense, the primary reason why locally convex spaces are so useful is that there are guaranteed to be plenty of continuous linear functionals, and the Hahn–Banach theorem provides them.

Theorem 3.9 (Hahn–Banach). *Suppose X is a vector space over $\mathbb{R}$, Y is a subspace, and $p : X \to \mathbb{R}$ is a gauge. Suppose $f : Y \to \mathbb{R}$ is a linear functional for which $f(y) \le p(y)$ for all $y \in Y$. Then f extends to a linear functional $F : X \to \mathbb{R}$ for which $F(x) \le p(x)$ for all $x \in X$. Finally, if X is a topological vector space and p is continuous, then F is continuous.*

Proof. We use Zorn's lemma. Consider the set $\mathscr{P}$ of all ordered pairs (g, Z), where Z is a subspace of X containing Y, and $g : Z \to \mathbb{R}$ is a linear functional that extends f and for which $g(z) \le p(z)$ for all $z \in Z$. Partially order $\mathscr{P}$ by extension: $(g, Z) \le (g', Z')$ when $Z \subset Z'$ and $g'|_Z = g$. $(f, Y) \in \mathscr{P}$, so $\mathscr{P}$ is nonempty. Zorn's lemma will produce a maximal element $(F, Y_0) \in \mathscr{P}$ once we know that every nonempty chain (i.e., totally ordered subset) in $\mathscr{P}$ is bounded.

Suppose $\mathscr{C}$ is a nonempty chain in $\mathscr{P}$. Set

$$Z_0 = \bigcup_{(g,Z)\in\mathscr{C}} Z \; ; \; g_0(x) = g(x) \text{ when } x \in Z.$$

The function g is well-defined, since $\mathscr{C}$ is a chain: if $x \in Z_1$ and $x \in Z_2$, with $(g_1, Z_1), (g_2, Z_2) \in \mathscr{C}$, then either $(g_1, Z_1) \leq (g_2, Z_2)$ [in which case $g_2|_{Z_1} = g_1$, so $x \in Z_1$ gives $g_1(x) = g_2(x)$], or $(g_2, Z_2) \leq (g_1, Z_1)$ (ditto, reversed). Also, Z_0 is a subspace, since $\mathscr{C}$ is a chain: if $x, y \in Z_0$, say $x \in Z_1, y \in Z_2$, with $(g_1, Z_1), (g_2, Z_2) \in \mathscr{C}$, then either $(g_1, Z_1) \leq (g_2, Z_2)$ (in which case $x \in Z_1 \subset Z_2$, so $x + y \in Z_2 \subset Z_0$), or $(g_2, Z_2) \leq (g_1, Z_1)$ (ditto, reversed). Also, if $x \in Z_0$ and $r \in \mathbb{R}$, say $x \in Z$ with $(g, Z) \in C$, then $rx \in Z \subset Z_0$. (This does not use the chain property.) Next, if $x \in Z_0$, say $x \in Z$ with $(g, Z) \in \mathscr{C}$, then $g_0(x) = g(x) \leq p(x)$. Finally, $Z_0 \supset Y$, and $g_0|_Y = f$, since $\mathscr{P}$ is nonempty: choosing any $(g, Z) \in \mathscr{P}: Y \subset Z \subset Z_0$, and for $y \in Y$, $g_0(y) = g(y) = f(y)$. So $(g_0, Z_0) \in \mathscr{P}$, and is, by construction, an upper bound for $\mathscr{C}$. Since $\mathscr{C}$ was arbitrary, $\mathscr{P}$ has a maximal element (F, Y_0). It remains to show that $Y_0 = X$.

Suppose $Y_0 \neq X$. Choose any $y_0 \notin Y_0$, and set $Z = Y_0 + \mathbb{R}y_0$. This sum is direct. The claim is that F can be extended to G, with $(G, Z) \in \mathscr{P}$ and $G|_{Y_0} = F$, contradicting maximality. To do this, set

$$G(y + ty_0) = F(y) + c_0 t$$

for a constant c_0 to be determined. This G certainly extends F to Z, and it remains only to show that c_0 can be chosen so that $G(x) \leq p(x)$ for all $x = y + ty_0 \in Z$. This holds by assumption (regardless of c_0) when $t = 0$. The final point is that the conditions required when $t > 0$ and when $t < 0$ are compatible.

Suppose $t > 0$. We need

$$F(y) + c_0 t \;=\; G(y + ty_0) \;\leq\; p(y + ty_0) = tp(t^{-1}y + y_0)$$
$$\|$$
$$t(F(t^{-1}y) + c_0) \qquad\qquad \text{that is (setting } y' = t^{-1}y)$$

$$c_0 \;\leq\; p(y' + y_0) \;-\; F(y') \text{ for all } y' \in Y_0.$$

Suppose $t < 0$. We need

$$F(y) + c_0 t = G(y + ty_0) \;\leq\; p(y + ty_0) = -tp((-t)^{-1}y - y_0)$$
$$\|$$
$$-t(F((-t)^{-1}y) - c_0) \qquad \text{that is (setting } y'' = (-t)^{-1}y)$$

$$-c_0 \leq p(y'' - y_0) - F(y''), \text{ or}$$

$$c_0 \geq -p(y'' - y_0) + F(y'') \text{ for all } y'' \in Y_0.$$

But: given any $y', y'' \in Y_0$,

$$F(y') + F(y'') = F(y' + y'') \leq p(y' + y'')$$
$$= p((y' + y_0) + (y'' - y_0)) \leq p(y' + y_0) + p(y'' - y_0)$$

that is $F(y'') - p(y'' - y_0) \leq p(y' + y_0) - F(y')$.

Set

$$c_0 = \inf_{y' \in Y_0} p(y' + y_0) - F(y').$$

Any $F(y'') - p(y'' - y_0)(y'' \in Y_0)$ is a lower bound for this set of values, so $c_0 \geq F(y'') - p(y'' - y_0)$ for all $y'' \in Y_0$ since c_0 is the *greatest* lower bound for these values. (In particular, $c_0 > -\infty$.)

Finally, $c_0 \leq p(y' + y_0) - F(y')$ for all $y' \in Y_0$, since c_0 is the greatest *lower bound* for these values.

There remains the question of continuity when X is a topological vector space. Suppose X is a topological vector space, and p is continuous. Given $\epsilon > 0$, there is an open neighborhood V of 0 for which $p(x) < \epsilon$ when $x \in V$, since $p(0) = p(0 \cdot 0) = 0 \cdot p(0) = 0$ from the gauge condition. If $x \in X$, and $y - x \in V \cap (-V)$, then $x - y \in V$ and $y - x \in V$, so

$$F(x) - F(y) = F(x - y) \leq p(x - y) < \epsilon \text{ and}$$

$$F(y) - F(x) = F(y - x) \leq p(y - x) < \epsilon, \text{ that is}$$

$$|F(x) - F(y)| < \epsilon \text{ when } y \in x + (V \cap (-V)).$$

$\square$

In the next section, this will be used to produce the topological dual for a Hausdorff locally convex space. There are some preliminary matters, however, concerning convex sets. Since our primary concern is with locally convex spaces and continuous linear functionals, we start with that picture.

Lemma 3.10. *Suppose X is a locally convex space, and suppose C is a closed, convex set, with $0 \in int(C)$. Suppose $x_0 \notin C$. Then there is a continuous linear functional $F : X \to \mathbb{R}$ for which $F(C) \subset (-\infty, 1]$ and $F(x_0) > 1$.*

Proof. Let p_C denote the Minkowski functional for C; p_C is a continuous gauge by Theorem 3.7. Since $p_C(x_0) < 1 \Rightarrow x_0 \in C$ (Theorem 3.7): $p_C(x_0) \geq 1$. If $p_C(x_0) = 1$, then $p_C\left(\left(1 - \frac{1}{n}\right) x_0\right) = 1 - \frac{1}{n} < 1 \Rightarrow \left(1 - \frac{1}{n}\right) x_0 \in C$, so $x_0 \in C$ since C is closed and $\left(1 - \frac{1}{n}\right) \to 1 \Rightarrow \left(1 - \frac{1}{n}\right) x_0 \to x_0$ (scalar multiplication is continuous). Hence $p_C(x_0) > 1$.

On $\mathbb{R}x_0$, set $f(tx_0) = tp_C(x_0)$. If $t \geq 0$, then $f(tx_0) = tp_C(x_0) = p_C(tx_0)$. If $t < 0$, then $f(tx_0) = tp_C(x_0) < 0 \leq p_C(tx_0)$. Hence $f(y) \leq p_C(y)$ for $y \in Y = \mathbb{R}x_0$. The Hahn–Banach theorem now extends f to a continuous linear functional $F : X \to \mathbb{R}$ for which $F(x) \leq p_C(x)$ for all $x \in X$. In particular, if $x \in C$, then $F(x) \leq p_C(x) \leq 1$, while $F(x_0) = f(x_0) = p_C(x_0) > 1$. $\square$

Corollary 3.11. *Suppose X is a locally convex space, and suppose C is a closed, convex set, with $0 \in C$. Suppose $x_0 \notin C$. Then there is a continuous linear functional $F : X \to \mathbb{R}$ for which $F(C) \subset (-\infty, 1]$ and $F(x_0) > 1$.*

Proof. Let $\mathcal{B}_0$ be a base for the topology at zero consisting of open, convex, balanced sets. Since C is closed,

$$x_0 \notin C = C^- = \bigcap_{W \in \mathcal{B}_0} C + W$$

so $x_0 \notin C + W$ for some $W \in \mathcal{B}_0$. Set

$$C' = \left(C + \frac{1}{2}W\right)^- \subset C + \frac{1}{2}W + \frac{1}{2}W \subset C + W.$$

Then $0 \in \frac{1}{2}W \subset C'$, so $0 \in \text{int}C'$. Also, $C \subset C'$. Finally, $x_0 \notin C'$ since $C' \subset C + W$. By Lemma 3.10, there is a continuous linear functional $F : X \to \mathbb{R}$ for which $F(x_0) > 1$ and $F(C) \subset F(C') \subset (-\infty, 1]$.   □

Proposition 3.12. *Suppose X is a locally convex space, and C_1 and C_2 are two disjoint nonempty convex sets, with C_1 closed and C_2 compact. Then there is a real number r_0 and a continuous linear functional $F : X \to \mathbb{R}$ for which $F(x) < r_0$ for all $x \in C_1$, and $F(x) > r_0$ for all $x \in C_2$.*

Proof. Pick any $y_0 \in C_1 - C_2$, a closed (Corollary 1.15) convex (Proposition 2.14 applied to $C = 2C_1$, $D = -2C_2$, and $I = [\frac{1}{2}, \frac{1}{2}]$) set. Then $0 \in (C_1 - C_2 - y_0)$. Also, $x_0 = -y_0 \notin C_1 - C_2 - y_0$ since $0 \notin C_1 - C_2$ (since $C_1 \cap C_2 = \emptyset$). Choose a continuous linear functional $F : X \to \mathbb{R}$ for which $F(x_0) > 1$ and $F(C_1 - C_2 - y_0) \subset (-\infty, 1]$. Note that if $x \in C_1$ and $y \in C_2$, then

$$1 \geq F(x - y - y_0) = F(x) - F(y) + F(-y_0)$$
$$= F(x) - F(y) + F(x_0), \text{ that is}$$
$$1 - F(x_0) \geq F(x) - F(y), \text{ that is}$$
$$F(y) - \frac{1}{2}(F(x_0) - 1) \geq \frac{1}{2}(F(x_0) - 1) + F(x).$$

Set (using compactness of C_2)

$$r_0 = \min_{y \in C_2} F(y) - \frac{1}{2}(F(x_0) - 1)$$

and observe that (with $\epsilon = \frac{1}{2}(F(x_0) - 1) > 0$) all $F(y) \geq r_0 + \epsilon$ $(y \in C_2)$ and all $F(x) \leq r_0 - \epsilon$ $(x \in C_1)$.   □

The preceding results give the main separation theorems for convex sets in locally convex spaces. The dual space, the set of continuous linear functionals with values in the base field, constitutes our next subject. Most of what is needed is a consequence of what appears in this section, with one (surmountable) complication coming from the possibility that our locally convex space is defined over the complex numbers.

3.4 The Dual

As noted earlier, the primary reason for assuming that a topological vector space is locally convex is to guarantee that there are enough continuous linear functionals to say something intelligent about the space. In order to evenly separate points, the space will also need to be Hausdorff, but there are a few cases where intermediate constructions yield non-Hausdorff locally convex spaces.

Definition 3.13. Suppose X is a locally convex topological vector space over $\mathbb{R}$ or $\mathbb{C}$. Letting $\mathbb{F}$ denote the base field, the **Dual space** of X, denoted by X^*, is the space of continuous linear functionals $f : X \to \mathbb{F}$. The **Algebraic Dual**, denoted by X', is the space of all linear functionals $f : X \to \mathbb{F}$, continuous or not.

Remarks. (1) The dual will be topologized (in more than one way!) in Sect. 3.6.
(2) The notation above for X^* and X' is fairly standard in books that use functional analysis in other fields. Unfortunately, it is the *opposite* of the most common usage in functional analysis textbooks. Be wary!

For locally convex spaces over $\mathbb{R}$, we have what we need, and the results of this section would only take a few more lines. For complex vector spaces, we must contend with the fact that the only linear functionals directly produced by the Hahn–Banach theorem are real-valued.

The following result makes the connection between real-linear functionals and complex-linear functionals, and it is not the least bit "obvious."

Proposition 3.14. *Suppose X is a locally convex topological vector space over $\mathbb{C}$, and $f : X \to \mathbb{R}$ is an $\mathbb{R}$-linear functional. Then there is a unique $\mathbb{C}$-linear functional $F : X \to \mathbb{C}$ for which $f(x) = \mathrm{Re}(F(x))$ for all $x \in X$. F is given by the formula $F(x) = f(x) - if(ix)$. If f is continuous, then so is F. Finally, if B is a convex, balanced subset of X for which $f(x) \le 1$ when $x \in B$, then $|F(x)| \le 1$ when $x \in B$.*

Proof. The formula starts things. First of all, if $f(x) = \mathrm{Re}(F(x))$ with F being $\mathbb{C}$-linear, then $f(ix) = \mathrm{Re}(F(ix)) = \mathrm{Re}(iF(x)) = -\mathrm{Im}(F(x))$, so our formula is forced. Set $F(x) = f(x) - if(ix)$. Then by inspection, $F : X \to \mathbb{C}$ is an $\mathbb{R}$-linear transformation, since $F(x) = f(x) \cdot 1 - f(ix) \cdot i$, and (1) multiplication by i on X is $\mathbb{R}$-linear, and (2) 1 and i are "vectors" in $\mathbb{C}$. With this definition, as well, F is continuous when f is, and $F(ix) = f(ix) - if(i^2x) = f(ix) - if(-x) = f(ix) + if(x)$, while $iF(x) = if(x) - i^2 f(ix) = if(x) + f(ix) = F(ix)$. Finally,

$$F((a + bi)x) = F(ax + ibx) = F(ax) + F(ibx)$$

$$= aF(x) + iF(bx) = aF(x) + ibF(x) = (a + bi)F(x),$$

so F is complex linear.

Now suppose B is a convex, balanced subset of X for which $f(x) \le 1$ when $x \in B$. Given $x \in B$, if $F(x) = |F(x)|e^{i\theta}$, then $e^{-i\theta}x \in B$ since B is balanced, and

$$|F(x)| = e^{-i\theta} F(x) = F(e^{-i\theta}x)$$

$$= \operatorname{Re}(F(e^{-i\theta}x)) \leq 1.$$

□

Corollary 3.15. *Suppose X is a locally convex topological vector space over $\mathbb{R}$ or $\mathbb{C}$, and suppose B is a nonempty closed, convex, balanced subset of X. If $x_0 \notin B$, then $\exists\, F \in X^*$ for which $|F(x)| \leq 1$ when $x \in B$, and $\operatorname{Re} F(x_0) > 1$. In particular, $|F(x_0)| > 1$.*

Proof. Since B is not empty, $\{0\} = 0 \cdot B \subset B$, that is $0 \in B$, so Corollary 3.11 applies to produce a continuous, $\mathbb{R}$-linear functional $f : X \to \mathbb{R}$ for which $f(x) \leq 1$ when $x \in B$, but $f(x_0) > 1$. If the base field is $\mathbb{C}$, then the $F \in X^*$ produced by Proposition 3.14 has the required properties by the last part of Proposition 3.14. If the base field is $\mathbb{R}$, simply note that $f(B)$ is a balanced subset of $\mathbb{R}$ [Proposition 2.17(b)] which is contained in $(-\infty, 1]$, so $f(B) \subset [-1, 1]$. That is, $|f(x)| \leq 1$ for all $x \in B$. In other words, if the base field is $\mathbb{R}$, we can directly take $F = f$.   □

In what follows, continuous linear functionals will usually be written in lower-case, since the preceding provides the main transition from $\mathbb{R}$ to $\mathbb{C}$. The next result is an exception.

Proposition 3.16. *Suppose X is a locally convex topological vector space, and Y is a subspace. Then any $f \in Y^*$ extends to some $F \in X^*$. That is, the restriction map $F \mapsto F\big|_Y$ from X^* to Y^* is onto.*

Proof. First case: Base field $= \mathbb{R}$. $\{x \in Y : f(x) < 1\}$ is a neighborhood of 0 in Y, and Y has the induced topology, so there is a convex, balanced neighborhood C of 0 in X such that $C \bigcap Y \subset \{x \in Y : f(x) < 1\}$. Thus, letting p_C denote the support function of C, $p_C(x) \geq f(x)$ for all $x \in Y$ by Proposition 3.8. Now the Hahn–Banach theorem extends f to $F \in X^*$, with $F(x) \leq p_C(x)$ for all $x \in X$. (F is continuous by the last part of the Hahn–Banach theorem, since p_C is continuous [Theorem 3.7].)

Second case: Base field $= \mathbb{C}$. Use the preceding to continuously extend $Re(f)$ to X, and use Proposition 3.14 to complexify that extension.   □

There is one last point to make before going on to the next section, in which Corollary 3.15 will provide the starting point:

Corollary 3.17. *Suppose X is a Hausdorff locally convex topological vector space over $\mathbb{R}$ or $\mathbb{C}$. Then X^* separates points. That is, if $x \neq y$, then $\exists\, f \in X^*$ for which $f(x) \neq f(y)$.*

Proof. Choose any such $f \in Y^*$, $Y = \operatorname{span}\{x, y\}$, and continuously extend it to X.   □

3.5 Polars

Corollary 3.15 provides the basic result for discussing polars, which have a number of uses. They will be used in the next section to topologize X^*, for example.

Definition 3.18. Suppose X is a locally convex topological vector space over $\mathbb{R}$ or $\mathbb{C}$. If $B \subset X$, then the **polar** of B, denoted by B°, is the set

$$\{f \in X^* : |f(x)| \leq 1 \,\forall\, x \in B\}.$$

If $A \subset X^*$, then the **polar** of A in X, denoted by $A_\circ$, is the set

$$\{x \in X : |f(x)| \leq 1 \,\forall\, f \in A\}.$$

Some (perhaps most) texts denote the polar of $A \subset X^*$ with the notation "A°." Since X^* will eventually be topologized as a local convex space, our A° will be in X^{**}, the dual of X^*, where it should be. The lower circle in $A_\circ$ is intended to indicate that $A_\circ$ is "down" in X. Again, be wary when reading the literature!

The next proposition describes the elementary properties that come directly from the definition.

Proposition 3.19. *Suppose X is a locally convex topological vector space over $\mathbb{R}$ or $\mathbb{C}$, and suppose $A, B \subset X$ and $D, E \subset X^*$. Then:*

(a1) $A \subset (A^\circ)_\circ$.
(a2) $D \subset (D_\circ)^\circ$.
(b1) $A \subset B \Rightarrow A^\circ \supset B^\circ$.
(b2) $D \subset E \Rightarrow D_\circ \supset E_\circ$.
(c1) $(A \cup B)^\circ = (A^\circ) \cap (B^\circ)$.
(c2) $(D \cup E)_\circ = (D_\circ) \cap (E_\circ)$.
(d1) *If $c \neq 0$, then $(cA)^\circ = c^{-1}(A^\circ)$.*
(d2) *If $c \neq 0$, then $(cD)_\circ = c^{-1}(D_\circ)$.*
(e) $A \subset D_\circ \Leftrightarrow A^\circ \supset D$.
(f) $D_\circ$ *is closed, convex, balanced, and nonempty.*

Proof. (a1) and (a2) come directly from the definition. So do (b1) and (b2), since stronger conditions yield smaller sets. For (c1),

$$(A \cup B)^\circ = \{f \in X^* : |f(x)| \leq 1 \text{ for } x \in A \cup B\}$$

$$= \{f \in X^* : |f(x)| \leq 1 \text{ for } x \in A \text{ or } x \in B\} = (A^\circ) \cap (B^\circ).$$

(c2) is similar. (d1) follows from the fact that

$$f \in (cA)^\circ \Leftrightarrow \forall\, x \in A(|cf(x)| = |f(cx)| \leq 1)$$

$$\Leftrightarrow cf \in A^\circ \Leftrightarrow f \in c^{-1}(A^\circ).$$

(d2) is similar. (e) follows from that fact that

$$A \subset D_\circ \Leftrightarrow \forall\, x \in A \,\forall\, f \in D(|f(x)| \le 1) \Leftrightarrow D \subset A^\circ.$$

Finally, (f) comes from the fact that

$$D_\circ = \bigcap_{f \in D} \{x \in X : |f(x)| \le 1\},$$

a closed, convex, balanced [Proposition 2.5(b)] set, with $0 \in D_\circ$.   □

Polars require some thought. They get smaller as the starting set gets larger, but do so in a regular way. They take a bit of getting used to, but it is worth the effort.

To proceed, here is a general definition, to be used routinely in what follows. Suppose X is a vector space over $\mathbb{R}$ or $\mathbb{C}$, and suppose $A, B \subset X$, with B being nonempty and balanced. We say B absorbs A if $A \subset cB$ for some scalar c. If $|d| \ge |c|$, then $c = db$ for $|b| \le 1$ (this is so even if $c = 0$), so $bB \subset B \Rightarrow cB = dbB \subset dB$. Hence $A \subset cB \Rightarrow A \subset dB$. Moral: $|c|$ only has to be large enough. (This is why this definition only works as is for B being balanced.)

Theorem 3.20 (Bipolar Theorem). *Suppose X is a locally convex topological vector space over $\mathbb{R}$ or $\mathbb{C}$; $A, B \subset X$; and $D \subset X^*$. Assume that B is closed, convex, balanced, and nonempty. Then:*

(a) $(B^\circ)_\circ = B$.
(b) $(A^\circ)_\circ$ is the smallest closed, convex, balanced, nonempty set containing A.
(c) B absorbs $A \Leftrightarrow B$ absorbs $(A^\circ)_\circ$.
(d) If A is bounded, then so is $(A^\circ)_\circ$.
(e) $D_\circ$ absorbs $A \Leftrightarrow A^\circ$ absorbs D.
(f) B absorbs $A \Leftrightarrow A^\circ$ absorbs B°.
(g) $((A^\circ)_\circ)^\circ = A^\circ$ and $((D_\circ)^\circ)_\circ = D_\circ$.

Proof. (a) Follows directly from Corollary 3.15 and Proposition 3.19(a1): $B \subset (B^\circ)_\circ$, but if $x_0 \notin B$, then $\exists\, f \in X^*$ for which $|f(x)| \le 1$ for all $x \in B$ but $|f(x_0)| > 1$. Hence this f belongs to B°, so that x_0 does not belong to $(B^\circ)_\circ$.

(b) $(A^\circ)_\circ$ is closed, convex, balanced, and nonempty [Proposition 3.19(f)], so suppose C is closed, convex, balanced, and nonempty, and $A \subset C$. Then $A^\circ \supset C^\circ$ [Proposition 3.19(b1)] and $(A^\circ)_\circ \subset (C^\circ)_\circ$ [Proposition 3.19(b2)]. But $C = (C^\circ)_\circ$ by part (a), so $(A^\circ)_\circ \subset C$.

(c) If $A \subset cB$, then (since cB is closed, convex, and balanced) $(A^\circ)_\circ \subset cB$ by part (b). If $(A^\circ)_\circ \subset cB$, then $A \subset cB$ since $A \subset (A^\circ)_\circ$ [Proposition 3.19(a2)].

(d) Suppose A is bounded. If U is a neighborhood of 0 in X, then there is a closed, convex, balanced neighborhood V of 0 with $V \subset U$ (Proposition 3.1). Then $A \subset cV$ for some scalar c. Hence $A^\circ \supset (cV)^\circ$ [Proposition 3.19(b1)] and $(A^\circ)_\circ \subset ((cV)^\circ)_\circ$ [Proposition 3.19(b2)]. But cV is closed, convex, and balanced, so $((cV)^\circ)_\circ = cV$ by part (a). Hence $(A^\circ)_\circ \subset cV \subset cU$.

(e) $A°$ *is* convex and balanced. If $c \neq 0$:

$$c(D_o) \supset A \Leftrightarrow (c^{-1}D)_o \supset A \qquad \text{(Proposition 3.19(d2))}$$

$$\Leftrightarrow c^{-1}D \subset A° \qquad \text{(Proposition 3.19(e))}$$

$$\Leftrightarrow D \subset c(A°) \qquad \text{(Multiplication by } c \text{ is bijective.)}$$

(f) By part (e) (with $D = B°$), $(B°)_o$ absorbs $A \Leftrightarrow A°$ absorbs $B°$. But $B = (B°)_o$ by part (a).

(g) $A \subset (A°)_o$ by Proposition 3.19(a1), so by Proposition 3.19(b1), with $B = (A°)_o$, we get $A° \supset ((A°)_o)°$. But $A° \subset ((A°)_o)°$ by Proposition 3.19(a2) (with $D = A°$). The fact that $((D_o)°)_o = D_o$ is similar: $D \subset (D_o)°$ by Proposition 3.19(a2), so $D_o \supset ((D_o)°)_o$ by Proposition 3.19(b2). But $D_o \subset ((D_o)°)_o$ by Proposition 3.19(a1) (with $A = D_o$).   □

One thing is worth contemplating. Part (a) is the crucial point; all else follows fairly directly, even though the style of argument is a bit unusual. Proposition 3.1 connects with our assumptions as well. Finally, part (a) itself is basically contained in Corollary 3.15, which comes from the Hahn–Banach theorem. The centrality of the Hahn–Banach theorem could hardly be clearer.

There is one last result, which relates compactness with polars.

Proposition 3.21. *Suppose X is a Hausdorff, locally convex topological vector space, and K is a compact, convex subset of X. Then $(K°)_o$ is compact.*

Case 1. Base field $= \mathbb{R}$: If $K = \emptyset$, then $(K°)_o = \{0\}$ by Theorem 3.20(b), since $\{0\}$ is the smallest nonempty closed, convex, balanced set. Suppose $K \neq \emptyset$. Set

$$E = \{tx + (1-t)y : x \in K, y \in -K, t \in [0,1]\}.$$

E is a continuous image of $K \times (-K) \times [0,1]$, a compact set, so E is compact, hence closed (X is Hausdorff). E is also convex (Proposition 2.14, with $I = [0,1]$). $0 \in E$ since $0 = \frac{1}{2} \cdot x + \left(1 - \frac{1}{2}\right)(-x)$ for $x \in K$ ($K \neq \emptyset$). If $0 \leq s \leq 1$, and $z \in E$, then $sz = sz + (1-s)0 \in E$ since E is convex. If $-1 \leq s \leq 0$ and $z \in E$, then $sz = -s(-z) \in E$ since $-z \in E$: if $z = tx + (1-t)y$, then

$$-z = (1-t)(-y) + t(-x) \in E.$$

So: E is closed, convex, balanced, and nonempty, so $(K°)_o \subset E$ by Theorem 3.20(b). As a closed subset of a compact set, $(K°)_o$ is compact.

Case 2. Base field $= \mathbb{C}$: As before, if $K = \emptyset$, then $(K°)_o = \{0\}$, so suppose $K \neq \emptyset$. Replace K with the set constructed in Case 1; without loss of generality, we may assume that K is also $\mathbb{R}$-balanced. Set

$$B = \{tx + (1-t)y : x \in K, y \in iK, t \in [0,1]\}.$$

As above, B is convex and compact, and B is $\mathbb{R}$-balanced since K is: if $-1 \le s \le 1$, then $s(tx + (1-t)y) = t(sx) + (1-t)(sy)$. Finally, set

$$C = \bigcap_{0 \le \theta < 2\pi} e^{i\theta}\sqrt{2}B.$$

C is convex and $\mathbb{C}$-balanced by Proposition 2.19 (with $\mathscr{F} = \{\{0\}\}$). C is also closed (it is an intersection of closed sets) and compact (it is a closed subset of the compact set $\sqrt{2}B$). $C \ne \emptyset$ since $0 \in B$. As above: $(K°)_\circ$ will be compact once we know that $(K°)_\circ \subset C$, which [via Theorem 3.20(b)] will hold once we know that $K \subset C$.

We need to show that $K \subset e^{i\theta}\sqrt{2}B$ for $0 \le \theta < 2\pi$, that is $e^{-i\theta}K \subset \sqrt{2}B$. Now the geometry comes in: $e^{-i\theta}$ is on the unit circle in $\mathbb{C}$, so $e^{-i\theta} = t(r) + (1-t)(is)$, for $-\sqrt{2} \le r \le \sqrt{2}$ and $-\sqrt{2} \le s \le \sqrt{2}$ since the convex hull of $[-\sqrt{2}, \sqrt{2}] \cup i[-\sqrt{2}, \sqrt{2}]$ includes the unit circle:

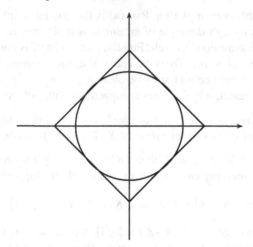

So, if $x \in K$, then

$$e^{i\theta}x = (tr + (1-t)is)x = \sqrt{2}\left(t\frac{r}{\sqrt{2}}x + (1-t)i\frac{s}{\sqrt{2}}x\right) \in \sqrt{2}B$$

since $\frac{r}{\sqrt{2}}x, \frac{s}{\sqrt{2}}x \in K$ (K is $\mathbb{R}$-balanced). $\qquad\qquad\qquad\qquad\qquad \square$

The next step is to use polars (and similar sets) to topologize spaces like X^*. This takes its own section.

3.6 Associated Topologies

At this point, we have all we need to construct the topologies (yes, plural!) on X^*, and most of what we need to define the topologies that come automatically with a locally convex space (or spaces). Unlike the next section, all but one of the constructions use polars and polar-like objects rather than seminorms (although seminorms appear in proofs). These topologies are produced via Theorem 3.2. We start with the two main topologies on X^*. (There will be nine [!] topologies, on various spaces, defined in this section: two on X^*, one on X', two on X, three on $\mathscr{L}_c(X, Y)$ and one on certain subspaces of X.)

Suppose X is a locally convex space. Using Theorem 3.2 to topologize X^* via polars, we need a family $\mathscr{B}_0$ of convex, balanced, absorbent sets. They will be polars of subsets of X, which automatically makes them convex and balanced. On the other hand, A° is absorbent if and only if, for all $f \in X^*$: there is a $c > 0$ such that $f \in c(A^\circ) = (c^{-1}A)^\circ$, that is for all $x \in A : c^{-1}|f(x)| = |f(c^{-1}x)| \leq 1$, or $|f(x)| \leq c$. This is guaranteed when A is bounded. (cf. Corollary 3.31 below).

The strong topology on X^*: It is defined by the family

$$\mathscr{B}_s = \{A^\circ : A \text{ is bounded in } X\}.$$

This works, since: (α) $\frac{1}{2}A^\circ = (2A)^\circ$; and (β) $A^\circ \cap B^\circ = (A \cup B)^\circ$, so that $\mathscr{B}_s$ is closed under intersections. This topology is also Hausdorff since $\cap \mathscr{B}_s = \{0\}$: if $f \in X^*$, and $f \neq 0$, so that $2 \in f(X)$ (since $f(X)$ is a nonzero subspace of the base field): $f(x) = 2 \Rightarrow f \notin \{x\}^\circ$.

The strong topology is the "default" topology on X^*, that is if X^* is referred to as a locally convex space with no further words as to which topology is used, then the topology is the strong topology. Often, for emphasis, X^* (with the strong topology) will be referred to as the "strong dual" of X.

The weak-* topology on X^*: It is defined by the family

$$\mathscr{B}_{w^*} = \{A^\circ : A \text{ is a finite subset of } X\}.$$

This works, since: (α) $\frac{1}{2}A^\circ = (2A)^\circ$; and (β) $A^\circ \cap B^\circ = (A \cup B)^\circ$, so that $\mathscr{B}_{w^*}$ is closed under intersections. This topology is also Hausdorff since $\cap \mathscr{B}_{w^*} = \{0\}$: if $f \in X^*$, and $f \neq 0$, so that $2 \in f(X)$ (since $f(X)$ is a nonzero subspace of the base field): $f(x) = 2 \Rightarrow f \notin \{x\}^\circ$.

(At this point, it should be clear that the same considerations keep coming up.) Observe that the weak-* topology is always Hausdorff, as is the (finer) strong topology, whether X is Hausdorff or not.

At times, in proofs, we will need to look at the *algebraic* dual, X', of X. The **weak-'** topology on X' is defined using the same mechanism: if A is a finite subset of X, set

$$A^{\circ'} = \{f \in X' : |f(x)| \leq 1 \text{ for all } x \in A\} \text{ and}$$

$$\mathscr{B}_{w'} = \{A^{\circ'} : A \text{ is a finite subset of } X\}.$$

This works for the same reasons as before. Note that, as a subspace of X', the subspace topology on X^* induced by the weak-' topology is the weak-* topology.

The weak topology on X: X already has a topology, which, following Rudin [32], we shall refer to as the *original* topology; but we define the weak topology on X by the family

$$\mathcal{B}_w = \{D_\circ : D \text{ is a finite subset of } X^*\}.$$

This works for the usual reasons. It is Hausdorff provided the original topology was Hausdorff, by Corollary 3.17.

At this point, we need some lemmas. For notation, we write X_w for X equipped with the weak topology, X_{w*}^* for X^* equipped with the weak-* topology and $X_{w'}'$ for X' equipped with the weak-' topology. (This pattern is fairly common in functional analysis.) Observe that practically by definition, every member of X^* is continuous on X_w, and evaluation at a point of X produces a continuous linear functional on X_{w*}^*. These are, in fact, the only ones; while the fact that $(X_w)^* = X^*$ can be deduced now (see Exercise 5), the other statement is less than clear, and both follow from this lemma:

Lemma 3.22. *Suppose X is a vector space over $\mathbb{R}$ or $\mathbb{C}$, and $f, f_1, \dots, f_n$ are linear functionals on X for which $\forall\, x \in X$:*

$$|f_1(x)| \le 1, \dots, |f_n(x)| \le 1 \Rightarrow |f(x)| \le 1.$$

Then $f = \Sigma c_j f_j$ for constants $c_1, \dots, c_n$.

Proof. Let $Y = \bigcap \ker f_j$. Note that $|f(x)| \le 1$ for all $x \in Y$. But $f(Y)$ is a subspace of $\mathbb{R}$ or $\mathbb{C}$, so $f(Y) = \{0\}$.

Now look on X/Y. Note that $\dim(X/Y) \le n$, and the linear functionals induced by $f_1, \dots, f_n$ span the algebraic dual of X/Y for linear-algebra reasons. (Think of X/Y as $\mathbb{R}^m$ or $\mathbb{C}^m$, and $f_1, \dots, f_n$ as rows of an $n \times m$ matrix. If its nullity is zero, then rank is m [rank-nullity theorem], so the rowspace consists of all $1 \times m$ matrices.) But f also induces a linear functional on X/Y, which is now a linear combination of those induced by $f_1, \dots, f_n$. $\qquad\square$

Corollary 3.23.

(a) $(X_w)^* = X^*$, as sets.
(b) Every continuous linear functional on X_{w*}^* is evaluation at a point of X.
(c) Every continuous linear functional on $X_{w'}'$ is evaluation at a point of X.

Proof. (b) [This part contains the main points.] Let $\mathbb{F}$ denote the base field, $\mathbb{R}$ or $\mathbb{C}$. Suppose $G \in (X_{w*}^*)^*$. Then $G^{-1}(\{c \in \mathbb{F} : |c| \le 1\}$ is a neighborhood of zero in X_{w*}^*, so it contains some A°, $A = \{x_1, \dots, x_n\} \subset X$. Hence

$$|f(x_1)| \le 1, \dots, |f(x_n)| \le 1 \Rightarrow f \in A^\circ \Rightarrow |G(f)| \le 1,$$

so that $G(f) = \Sigma c_j f(x_j)$ by Lemma 3.22. Hence $G(f) = f(\Sigma c_j x_j)$.

(a) If $f \in X'$, and $\{x \in X : |f(x)| \le 1\}$ contains $D_\circ$, $D = \{f_1, \ldots, f_n\} \subset X^*$, then (as above) $f = \Sigma c_j f_j \in X^*$.

(c) Just like (b), except the point in X actually is unique even if X is not Hausdorff, since X' does separate points. □

The next point comes from an observation. Again, letting $\mathbb{F}$ denote the base field, $\mathbb{R}$ or $\mathbb{C}$, X^* and X' consist of $\mathbb{F}$-valued functions on X. That is, quite literally:

$$X^* \subset X' \subset \prod_{x \in X} \mathbb{F}.$$

Proposition 3.24. *The* weak-* *topology on X^*, and the* weak-' *topology on X', are their subspace topologies of the product topology on $\prod_{x \in X} \mathbb{F}$.*

Proof. We need only work with X', since the weak-' topology on X' induces the weak-* topology on X^*.

Every product neighborhood of zero contains a weak-' neighborhood of zero:

Suppose $f \in X'$, and $|f(x_1)| < \epsilon_1, \ldots, |f(x_n)| < \epsilon_n$, a typical requirement for a neighborhood of zero in $\prod \mathbb{F}$. This can be enforced by requiring that

$$\left| f\left(\frac{2}{\epsilon_1} x_1\right) \right| \le 1, \ldots, \left| f\left(\frac{2}{\epsilon_n} x_n\right) \right| \le 1, \text{ that is}$$

$$f \in \left\{ \frac{2}{\epsilon_1} x_1, \ldots, \frac{2}{\epsilon_n} x_n \right\}^{\circ'}.$$

Every standard weak-' neighborhood of zero is the intersection of X' with a product neighborhood of zero:

If $A = \{x_1, \ldots, x_n\}$, then $f \in A^{\circ'}$ when every $|f(x_j)| \le 1$, and that defines a (closed) product neighborhood. □

We need to say more to make use of this.

Proposition 3.25. *X' is a closed subspace of $\prod_{x \in X} \mathbb{F}$.*

Proof. X' is the intersection of the kernels of all the following continuous linear functionals on $\prod_{x \in X} \mathbb{F}$: For $f \in \prod_{x \in X} \mathbb{F}$,

$$\forall x, y \in X : f \mapsto f(x) + f(y) - f(x + y)$$

$$\forall x \in X, \forall c \in \mathbb{F} : f \mapsto cf(x) - f(cx).$$

All these are made up of coordinate evaluations, and so are continuous.

□

By the way, X^* is *not* necessarily closed in $\prod \mathbb{F}$. In fact, X^* is dense in X' in the weak-' topology; see Exercise 4.

Theorem 3.26 (Banach–Alaoglu). *Suppose X is a locally convex space, and U is a neighborhood of zero in X. Then U° is weak-* compact.*

Proof. Let C be a closed, convex, balanced neighborhood of 0 contained in U, and let p_C denote its support function. p_C is a continuous seminorm by Theorem 3.7. If $f \in X'$ and $|f(x)| \le p_C(x)$ for all x, then $f \in X^*$ by Theorem 3.9. Finally, if $f \in U^\circ$, then for all $x \in X$ (cf. Proposition 3.3):

$$\forall \epsilon > 0 : p_C(x) + \epsilon \in I_x \Rightarrow x \in (p_C(x) + \epsilon)C \subset (p_C(x) + \epsilon)U$$

$$\Rightarrow \frac{1}{p_C(x) + \epsilon} x \in U \Rightarrow \left| f\left(\frac{1}{p_C(x) + \epsilon} x\right)\right| \le 1$$

$$\Rightarrow |f(x)| \le p_C(x) + \epsilon.$$

Since this holds for all $\epsilon > 0$, $f \in U^\circ \Rightarrow |f(x)| \le p_C(x)$. Hence U° is a weak-* closed (by its definition) subset of the compact (Theorem A.16) set

$$X' \cap \prod_{x \in X} \{c \in \mathbb{F} : |c| \le p_C(x)\}. \qquad \square$$

We can now define our second new topology on X.

The Mackey topology on X: It is defined by the family

$$\mathscr{B}_M = \{D_\circ : D \text{ is convex and weak-* compact in } X^*\}.$$

This works for the usual reasons, with one major glitch: $D_\circ \cap E_\circ = (D \bigcup E)_\circ$, but $D \bigcup E$ is not convex. However,

$$G = \{tf + (1 - t)g : f \in D, g \in E, t \in [0, 1]\}$$

is compact (it is a continuous image of $D \times E \times [0, 1]$) and convex (Proposition 2.14), and $G_\circ \subset D_\circ \cap E_\circ$. Observe that by the bipolar theorem (applied to X^*_{w*}) and Proposition 3.21, we can replace D with $(D_\circ)^\circ$, so that:

$$\mathscr{B}_M = \{D_\circ : D \text{ is weak-* compact in } X^* \text{ and } D = (D_\circ)^\circ\}.$$

Finally, while the weak topology is coarser than the original topology, the Mackey topology is finer by Banach–Alaoglu: if U is an original neighborhood of 0, and V is a closed, convex, balanced neighborhood of 0 contained in U, then $V = (V^\circ)_\circ$ (bipolar theorem) while V° is weak-* compact, so

$$U \supset V = (V^\circ)_\circ \in \mathscr{B}_M.$$

Now any member of X^* is continuous in the original topology, hence is continuous in the finer Mackey topology. In fact, we should pick up more continuous linear functionals, but we don't.

Proposition 3.27. *If X is a locally convex space, then any $f \in X'$ which is continuous when X is equipped with the Mackey topology, belongs to X^*.*

Proof. Suppose $f \in X'$, and f is Mackey-topology-continuous. Then $\{x \in X : |f(x)| \leq 1\}$ contains a Mackey neighborhood of zero, $D_\circ$, with $D = (D_\circ)^\circ$, and D being weak-* compact, convex, and balanced. (All polars A° are convex and balanced.) Look at D as a subset of X' with the weak-' topology: *In that context,* D is compact (hence closed in $X'_{w'}$!), convex, and balanced; so $D = (D^\circ)_\circ$ by the Bipolar theorem applied to $X'_{w'}$. But by Corollary 3.23(c), the dual of $X'_{w'}$ "is" X, so that

$$D^\circ \text{ in } (X'_{w'})^* \leftrightarrow D_\circ \text{ in } X$$

$$(D^\circ)_\circ \text{ in } X'_{w'} \leftrightarrow (D_\circ)^{\circ'} \text{ in } X'$$

so $D = (D_\circ)^{\circ'}$.

But $\{x \in X : |f(x)| \leq 1\} \supset D_\circ$, so $|f(x)| \leq 1$ for all $x \in D_\circ$. Hence $f \in (D_\circ)^{\circ'} = D \subset X^*$. $\square$

So, we started with a locally convex topology on X, and produced the weak topology (which was weaker) and the Mackey topology (which was stronger). Both were really constructed from X as a vector space, with the *only* relevant datum being that X^* was its dual space. Now freeze X^* and let the original topology float: there are various locally convex topologies on X for which X^* is the dual. The finest is the Mackey topology, while the coarsest is the weak topology. In many applications, that finer topology arises automatically (cf. Sect. 4.1).

Definition 3.28. Suppose X is a Hausdorff locally convex space. X is called a **Mackey space** if the original topology on X agrees with the Mackey topology.

There are a few final notes concerning these topologies, before going on to spaces of linear transformations. The first is basically an observation, but it deserves isolation as a theorem:

Theorem 3.29. *Suppose X is a locally convex space, and C is an originally closed, convex subset of X. Then C is weakly closed.*

Proof. Proposition 3.12 (along with Proposition 3.14 if the base field is $\mathbb{C}$) provides a means of separating C from any $x \notin C$, by setting $C_1 = C, C_2 = \{x\} : \{y \in X : F(y) > r_0\}$ is a weakly open neighborhood of x in $X - C$. $\square$

In particular, the original topology is locally weakly closed. Continuing in this manner, the strong topology on X^* is locally weak-* closed. In fact, the strong topology on X^* is the finest locally convex topology on X^* which is locally weak-* closed, cf. Exercise 7. While we have no application for this, it is illuminating.

Now, suppose X and Y are two locally convex spaces. There remain the three topologies on $\mathscr{L}_c(X, Y)$—the space of continuous linear transformations from X to Y.

The topology of bounded convergence: If A is bounded in X, and U is a convex, balanced neighborhood of 0 in Y, set $N(A, U) = \{T \in \mathscr{L}_c(X, Y) : T(A) \subset U\}$. The topology of bounded convergence is defined by

$$\mathscr{B}_b = \{N(A, U) : A \text{ is bounded in } X \text{ and } U \text{ is}$$

$$\text{a convex, balanced neighborhood of 0 in } Y\}.$$

Note that if $T \in \mathscr{L}_c(X, Y)$, then for A bounded in X, $T(A)$ is bounded in Y, so that if U is a convex, balanced neighborhood of 0 in Y, then $\exists c > 0$ for which $T(A) \subset cU$, that is $c^{-1}T(A) \subset U$, that is $c^{-1}T \in N(A, U)$, that is $T \in cN(A, U)$. Hence each $N(A, U)$ is absorbent. Also, $\frac{1}{2}N(A, U) = N(2A, U)$, while $N(A, U) \cap N(B, V) \supset N(A \cup B, U \cap V)$, so Theorem 3.2 works here as well.

The topology of pointwise convergence: It is defined by the family

$$\mathscr{B}_p = \{N(A, U) : A \text{ is finite in } X \text{ and } U \text{ is}$$

$$\text{a convex, balanced neighborhood of 0 in } Y\}.$$

By the way, especially when $X = Y$ and X is a Banach space, the topology of pointwise convergence goes by the totally confusing title of the *strong operator topology*. There is also a *weak operator topology*, which is just the topology of pointwise convergence on $\mathscr{L}_c(X, Y_w)$, restricted to $\mathscr{L}_c(X, Y)$. Observe that these topologies can also be applied to the space of bounded linear transformations from X to Y. Also, the topology of bounded convergence on $\mathscr{L}_c(X, \mathbb{F})$ is just the strong topology (Exercise 8).

Now for that subspace topology. For later clarity, the space will be Y.

Proposition 3.30. *Suppose Y is a Hausdorff locally convex space, and suppose D is a nonempty, bounded, closed, convex, balanced subset of Y. Then the domain of the Minkowski functional p_D:*

$$Y_D = \bigcup_{r>0} rD$$

is a subspace of Y, and (Y_D, p_D) is a normed space with the closed unit ball D. Furthermore, the norm topology on Y_D is finer than the topology induced from Y, so that any $f \in Y^*$ restricts to a norm-continuous linear functional on Y_D. Finally, if D is sequentially complete as a subset of the topological group $(Y, +)$, then (Y_D, p_D) is a Banach space.

Proof. $rD + sD = (r + s)D$ by Proposition 3.3, so D is closed under addition. It is closed under multiplication by any $\pm r$ (respectively any $re^{i\theta}$) when the base field is $\mathbb{R}$ (respectively $\mathbb{C}$), so Y_D is a subspace. The Minkowski functional p_D is a seminorm by Theorem 3.7.

If $x \neq 0$, $x \in Y_D$, then $\exists$ a convex, balanced neighborhood U of 0 with $x \notin U$ since Y is Hausdorff. $\exists \, r > 0$ with $D \subset rU$ since D is bounded, so that $r^{-1}D \subset U$, giving $x \notin r^{-1}D$ and $p_D(x) \geq r^{-1} > 0$. Thus, p_D is actually a norm.

Now, if $x \in D$, then $p_D(x) \leq 1$. But if $p_D(x) \leq 1$, then

$$p_D\left(\left(1 - \frac{1}{n}\right)x\right) \leq 1 - \frac{1}{n} < 1$$

$$\Rightarrow \left(1 - \frac{1}{n}\right)x \in D \qquad \text{(Theorem 3.7)}$$

$$\Rightarrow x = \lim_{n \to \infty}\left(1 - \frac{1}{n}\right)x \in D \qquad \text{(D is closed).}$$

Thus, D is the closed unit ball in (Y_D, p_D).

If U is any convex balanced neighborhood of 0 in Y, then the fact that $D \subset rU$ for some $r > 0$ shows that $r^{-1}D \subset U \cap Y_D$, so that the induced-topology neighborhood $U \cap Y_D$ of 0 contains a p_D-neighborhood, so the norm topology is finer than the induced topology.

Finally, suppose D is sequentially complete. Let $\langle x_n \rangle$ be a Cauchy sequence in (Y_D, p_D). If U is a neighborhood of 0 in Y, then $D \subset rU$ for some $r > 0$ since D is bounded. If $p_D(x) \leq r^{-1}$, then $p_D(rx) \leq 1$, so $rx \in D \subset rU$, that is $x \in U$. Thus, $p_D(x_n - x_m) \leq r^{-1} \Rightarrow x_n - x_m \in U$. This can be forced by making m and n large, so $\langle x_n \rangle$ is Cauchy in Y. Now $\langle x_n \rangle$ is bounded in (Y_D, p_D), so $\exists \, s > 0$ such that all $x_n \in sD$. But now $s^{-1}x_n \in D$ for all n, and $\langle s^{-1}x_n \rangle$ is Cauchy in Y, $(x_n - x_m \in sU \Rightarrow s^{-1}x_n - s^{-1}x_m \in U)$, so $s^{-1}x_n \to y$ for some $y \in D$ since D is sequentially complete. Hence $x_n \to sy$ in Y since multiplication by s is continuous. It remains to show that $x_n \to sy$ in (Y_D, p_D).

Suppose $\varepsilon > 0$. Choose N so that $m, n \geq N \Rightarrow p_D(x_m - x_n) \leq \varepsilon$. Then for $m, n \geq N$, $p_D(\varepsilon^{-1}(x_n - x_m)) \leq 1$, so that $\varepsilon^{-1}(x_n - x_m) \in D$, that is $x_n - x_m \in \varepsilon D$. Now freeze m and let $n \to \infty$. Then $x_n - x_m \to sy - x_m$ in Y, and εD is closed in Y, so $sy - x_m \in \varepsilon D$ when $m \geq N$. That is, $\varepsilon^{-1}(sy - x_m) \in D$, so that $p_D(\varepsilon^{-1}(sy - x_m)) \leq 1$, or $p_D(sy - x_m) \leq \varepsilon$, when $m \geq N$. That gives norm convergence. $\qquad \square$

Corollary 3.31. *Suppose X is a locally convex space, and $A \subset X$. The following are equivalent:*

(i) *A is originally bounded, that is A is bounded when X is equipped with the original topology.*

(ii) *A is weakly bounded, that is A is bounded when X is equipped with the weak topology.*

(iii) *Every $f \in X^*$ is bounded on A.*

(iv) *A° is absorbent.*

(v) *Every continuous seminorm p on X is bounded on A.*

Proof. (i) $\Rightarrow$ (ii) If A is absorbed by all original neighborhoods of zero, then A is absorbed by weak neighborhoods of zero, since weak neighborhoods of zero are original neighborhoods of zero.

(ii) $\Rightarrow$ (iv) If A is absorbed by all weak neighborhoods of zero, then $\forall f \in X^*$: $\{f\}_\circ$ absorbs A, so that A° absorbs $\{f\}$ [Theorem 3.20(e)]. That is, A° is absorbent.

(iv) $\Rightarrow$ (iii) If $f \in cA^\circ$, $c > 0$, then $c^{-1} f \in A^\circ$, so that $|c^{-1} f(x)| \leq 1$ for all $x \in A$, that is $|f(x)| \leq c$ for all $x \in A$.

(i) $\Rightarrow$ (v) $U = \{x \in X : p(x) \leq 1\}$ is a neighborhood of 0, and $A \subset rU$ ($r \in \mathbb{R}, r > 0$) $\Rightarrow p(x) \leq r$ for all $x \in A$.

(v) $\Rightarrow$ (iii) $|f|$ is a seminorm. Finally,

(iii) $\Rightarrow$ (i) Let Y denote X^* equipped with the weak-* topology, and suppose U is a closed, convex, balanced neighborhood of 0 in X. Set $D = U^\circ$. D is nonempty, closed, convex, and balanced: All polars are. D is also weak-* compact in Y by the Banach–Alaoglu theorem, hence is complete in $(Y, +)$ by Corollary 1.32, hence is sequentially complete, so that (Y_D, p_D) is a Banach space. If $x \in A$, let E_x denote evaluation at x; by assuming (iii), we are assuming that the set $\{E_x\}$ is a pointwise bounded family of continuous linear functionals on (Y_D, p_D), so that $\{E_x\}$ is operator-norm bounded by the Banach–Steinhaus theorem (a.k.a. the uniform boundedness principle) for Banach spaces. That is, $\exists M$ such that $\forall x \in A : |E_x(f)| \leq Mp_D(f)$ for $f \in Y_D$. In other words, $\forall x \in A : |f(x)| \leq Mp_D(f)$, so that $f \in D \Rightarrow |f(x)| \leq M \Rightarrow |f(M^{-1}x)| \leq 1$. But *that* just means that $M^{-1}x \in D_\circ = (U^\circ)_\circ = U$ since U is closed, convex, and balanced. Hence $M^{-1}A \subset U$, so $A \subset MU$. $\square$

The preceding is *very* important, and is typical of how the Banach–Alaoglu theorem and some auxiliary constructions such as (Y_D, p_D) arise in practical situations. Also, the completeness argument in Proposition 3.30 illustrates the utility of having a topology (in this case the norm topology on (Y_D, p_D)) which is locally closed in a coarser topological space Y. Proposition 3.30 will be used in Sects. 3.8, 4.1, and 6.1; it is quite handy at times. Finally, we now know why the sets A with polars that were taken to form the strong topology on X^* were originally bounded, while they "only" had to be weakly bounded to make their polars absorbent: the two notions of "bounded" coincide.

3.7 Seminorms and Fréchet Spaces

If X is a vector space over $\mathbb{R}$ or $\mathbb{C}$, and $p : X \to \mathbb{R}$ is a norm, then (X, p) is a normed space with a well-defined topology given by the metric $d(x, y) = p(x - y)$. The resulting space is Hausdorff and locally convex. If p is only a seminorm, then one can still make X into a locally convex space; it just will not be Hausdorff. How about a whole family of seminorms? You can still use the family to make a locally convex space, and this is how some spaces are most naturally defined.

For technical reasons, it pays to start the process with families that are directed. If p_1 and p_2 are seminorms on a vector space X over $\mathbb{R}$ or $\mathbb{C}$, then $p_1 \leq p_2$ when $p_1(x) \leq p_2(x)$ for all $x \in X$. A family $\mathscr{F}$ of seminorms is *directed* if for all $p_1, p_2 \in \mathscr{F}$, there exists $p_3 \in \mathscr{F}$ for which $p_1 \leq p_3$ and $p_2 \leq p_3$. Note that if one sets, for $p \in \mathscr{F}$ and $0 < r < \infty$,

$$B(p,r) = \{x \in X : p(x) < r\},$$

then

$$p_1 \leq p_2 \text{ and } r \geq s \Rightarrow B(p_1,r) \supset B(p_2,s),$$

$$\text{since } p_2(x) < s \Rightarrow p_1(x) \leq p_2(x) < s \leq r.$$

Definition 3.32. Suppose X is a vector space over $\mathbb{R}$ or $\mathbb{C}$, and $\mathscr{F}$ is a directed family of seminorms on X. Set

$$\mathscr{B}_0 = \{B(p,2^{-n}) : p \in \mathscr{F}, n \in \mathbb{N}\}.$$

Then $\mathscr{B}_0$ satisfies the conditions given in Theorem 3.2:

$$B(p_3,2^{-\ell}) \subset B(p_1,2^{-m}) \bigcap B(p_2,2^{-n})$$

if $p_1 \leq p_3$, $p_2 \leq p_3$, and $\ell = \max(m,n)$. The **topology induced by** $\mathscr{F}$ on X is the topology produced by Theorem 3.2, that is the topology having $\mathscr{B}_0$ as a base at 0.

Several things are worth noting. First of all, if $\mathscr{F}$ is countable or finite, then $\mathscr{B}_0$ is countable. Second, the procedure can be reversed:

Proposition 3.33. *Suppose X is a locally convex space over $\mathbb{R}$ or $\mathbb{C}$. Let $\mathscr{B}_1$ be a base at 0 for the topology consisting of convex, balanced sets. Set*

$$\mathscr{F} = \{p_V : V \in \mathscr{B}_1\}.$$

where p_V is the Minkowski functional associated with V, as described in Sect. 3.2. Then $\mathscr{F}$ is a directed family of seminorms on X, and the topology on X induced by $\mathscr{F}$ is the original topology on X.

Proof. Each p_V is a seminorm (Theorem 3.7), and if $U, V \in \mathscr{B}_1$, $\exists W \in \mathscr{B}_1$ with $W \subset U \bigcap V$. But, for example, $W \subset U \Rightarrow (x \in tW \Rightarrow x \in tU)$, so that $\forall x \in X$,

$$\{t \geq 0 : x \in tW\} \subset \{t \geq 0 : x \in tU\}, \text{ so that}$$

$$p_W(x) \geq p_U(x),$$

by looking at the left endpoints of the corresponding intervals. It follows that $W \subset U \cap V \Rightarrow p_W \geq p_U$ and $p_W \geq p_V$. Hence $\mathscr{F}$ is directed.

Each $p_V \in \mathscr{F}$ is continuous in the original topology (Theorem 3.7), so all elements of $\mathscr{B}_0$ are open in the original topology. Since $\mathscr{B}_1$ is a base for the original topology:

$$\forall\, U \in \mathscr{B}_0 \,\exists\, V \in \mathscr{B}_1 \text{ s.t. } V \subset U. \tag{$*$}$$

But if $V \in \mathscr{B}_1$, then $B(p_V, 1) \subset V$ by Theorem 3.7, so

$$\forall\, V \in \mathscr{B}_1 \,\exists\, U \in \mathscr{B}_0 \text{ s.t. } U \subset V. \tag{$**$}$$

Combining $(*)$ and $(**)$, a subset A of X contains a member of $\mathscr{B}_0$ if and only if it contains a member of $\mathscr{B}_1$, so $\mathscr{B}_0$ and $\mathscr{B}_1$ give the same answer to the question, "Is 0 interior to A?" (See step 2 in the proof of Proposition 1.8.) Since both topologies are translation invariant, and produce the same interiors, they are the same. □

Corollary 3.34. *Suppose X is a locally convex space over $\mathbb{R}$ or $\mathbb{C}$. Then the topology of X is given by a directed family of seminorms. This family can be chosen to be countable if X is first countable.*

Proof. $\mathscr{B}_1$ exists by Proposition 3.1. □

By the way, Reed and Simon [29] define a locally convex space this way.

What does one do if the family is not directed? There is a standard construction that goes as follows, if $\mathscr{F}_0$ is *any* family of seminorms.

1. If $\mathscr{F}_0$ is finite, set $\mathscr{F} = \left\{ \sum_{p \in \mathscr{F}_0} p \right\}$.

2. If $\mathscr{F}_0$ is countably infinite, write $\mathscr{F}_0 = \{p_1, p_2, p_3, \ldots\}$, and set

$$\mathscr{F} = \left\{ \sum_{j=1}^{n} p_j : n = 1, 2, \ldots \right\}$$
$$= \{p_1, p_1 + p_2, p_1 + p_2 + p_3, \ldots\}.$$

3. If $\mathscr{F}_0$ is uncountable, set

$$\mathscr{F} = \left\{ \sum_{p \in F} p : F \text{ is a finite subset of } \mathscr{F}_0 \right\}.$$

Suppose $x_\alpha \to x$ in the $\mathscr{F}$-topology, where $\langle x_\alpha \rangle$ is a net, and $\mathscr{F}$ is defined above. Then $p(x_\alpha - x) \to 0$ in $\mathbb{R}$ for all $p \in \mathscr{F}$ since each $p \in \mathscr{F}$ is continuous. Hence $p(x_\alpha - x) \to 0$ for all $p \in \mathscr{F}_0$ by squeezing. On the other hand, if $\langle x_\alpha \rangle$ is a net in X, and $x \in X$, and $p(x_\alpha - x) \to 0$ for all $p \in \mathscr{F}_0$, then $p(x_\alpha - x) \to 0$ for all $p \in \mathscr{F}$

(finite sums), so that for all $n \in \mathbb{N}$, there exists β such that $\alpha \succ \beta \Rightarrow p(x_\alpha - x) < 2^{-n}$, that is $x_\alpha \in x + B(p, 2^{-n})$. That is, $x_\alpha \to x$ in the topology induced by $\mathscr{F}$ if and only if $p(x_\alpha - x) \to 0$ for all $p \in \mathscr{F}_0$. In particular, the convergent nets [and thus the topology, by Proposition 1.3(a)] does not depend on the ordering of the seminorms in Case 2 above.

Now suppose $\mathscr{F} = \{p_1, p_2, \ldots\}$ is a countable (ascending) sequence of seminorms on X. For $x, y \in X$, set

$$d(x, y) = \sum_{j=1}^{\infty} 2^{-j} \frac{p_j(x - y)}{1 + p_j(x - y)}.$$

For the usual reasons, this defines a metric on X provided $\mathscr{F}$ is *separating*, that is $x \neq 0 \Rightarrow p(x) > 0$ for some $p \in \mathscr{F}$. The triangle inequality holds because $a, b \geq 0$ gives

$$\int_b^{a+b} \frac{dx}{(1+x)^2} = \int_0^a \frac{dx}{(1+b+x)^2} \leq \int_0^a \frac{dx}{(1+x)^2},$$

$$\text{that is } \frac{a+b}{1+a+b} - \frac{b}{1+b} \leq \frac{a}{1+a}.$$

This metric is translation invariant as well. Finally, note that if $x_n \to x$ in the metric topology, then every $p_j(x_n - x) \to 0$ (squeezing), while if $x_n \to x$ in the $\mathscr{F}$-topology, then every $p_j(x_n - x) \to 0$, so that $d(x_n, x) \to 0$ by the Lebesgue dominated convergence theorem for integrals (i.e., sums) over the positive integers (dominating function 2^{-j}). Thus, the metric gives the $\mathscr{F}$-topology. We have (nearly) proved:

Theorem 3.35. *Suppose X is a Hausdorff locally convex space. Then the following are equivalent:*

(i) *X is first countable.*
(ii) *X is metrizable.*
(iii) *The topology of X is given by a translation invariant metric.*
(iv) *The topology of X is given by a countable family of seminorms.*

Proof. The earlier discussion gives (iv) $\Rightarrow$ (iii). The implications (iii) $\Rightarrow$ (ii) and (ii) $\Rightarrow$ (i) are direct, while (i) $\Rightarrow$ (iv) comes from Corollary 3.34. $\square$

Next, a few words about completeness. A Hausdorff locally convex space (for that matter, a Hausdorff topological vector space) X is called *complete* (respectively, *sequentially complete*) if the additive topological group $(X, +)$ is complete (respectively, sequentially complete) as a topological group. Completeness and sequential completeness for subsets also refers to $(X, +)$ as a topological group.

Corollary 3.36. *Suppose X is a Hausdorff locally convex space. Then the following are equivalent:*

(i) X is first countable and complete.

(ii) X is metrizable and complete.

(iii) The topology of X is given by a complete, translation invariant metric.

(iv) X is complete, and the topology of X is given by a countable family of seminorms.

Proof. Thanks to Theorem 3.35, the only issue is the variation in "completeness" in condition (iii). Sequences are all we need to consider, thanks to Theorem 1.34.

The idea is this: A sequence $\langle x_n \rangle$ is Cauchy in the locally convex topology exactly when we can force $d(x_n - x_m, 0) < \varepsilon$ by requiring both n and m to be large. But

$$d(x_n - x_m, 0) = d(x_n - x_m + x_m, 0 + x_m) = d(x_n, x_m)$$

since d is translation invariant. That is, d and $(X, +)$ have the same Cauchy sequences (as well as the same convergent sequences), so if one is complete, then so is the other.   □

Definition 3.37. A **Fréchet space** is a Hausdorff locally convex space satisfying any (hence all) of conditions (i)–(iv) in Corollary 3.36.

By the way, for historical reasons (mainly Bourbaki [5]), a Fréchet space is usually defined using condition (ii). When reading condition (ii), keep in mind that "complete" really refers to X as a locally convex space, not to the metric appearing in "metrizable." It is only for translation invariant metrics that one can identify metric-Cauchy sequences with topological group-Cauchy sequences.

Examples of Fréchet Spaces

I. $\mathbb{R}[[x]]$ and $\mathbb{C}[[x]]$. (Formal power series.) The nth seminorm of $\sum a_n x^n$ is $|a_n|$, or $\sum_{i=0}^{n} |a_j|$ once these are transformed into a directed set. This is one of the simplest, yet it illustrates a complication with the earlier constructions. The metric gives

$$d(\sum a_n x^n, 0) = \sum_{n=0}^{\infty} 2^{-n} \frac{\sum_{j=0}^{n} |a_j|}{1 + \sum_{j=0}^{n} |a_j|}.$$

With this metric, *the ball of radius r need not be convex!*

Example. $r = 1.4$, $f(x) = 2$, and $g(x) = 16x$

$$d(2, 0) = \sum_{n=0}^{\infty} 2^{-n} \frac{2}{3} = \frac{4}{3} < 1.4$$

$$d(16x, 0) = \sum_{n=1}^{\infty} 2^{-n} \frac{16}{17} = \frac{16}{17} < 1.4$$

$$d(1 + 8x, 0) = \frac{1}{2} + \sum_{n=1}^{\infty} 2^{-n} \cdot \frac{9}{10} = 1.4$$

There is a way to get around this—replace all those sums earlier in this section with maxima. Rudin [32] does this. The cost is that some arguments become complicated due to the unavailability of convergence theorems for sums (i.e., integrals) over the positive integers.

II. $C(H)$, the continuous functions on a locally compact, σ-compact Hausdorff space. H can be written as

$$H = \bigcup_{n=1}^{\infty} K_n$$

where each K_n is compact, and each $K_n \subset \text{int}(K_{n+1})$. Set

$$p_n(f) = \max |f(K_n)|.$$

Fréchet space convergence is uniform convergence on compact sets.

III. $\mathscr{H}(U)$, the space of holomorphic functions on a region $U \subset \mathbb{C}$. The topology here comes from $C(U)$, via Example II.

IV. $C^{\infty}(\mathbb{R}^n)$, the space of C^{∞} functions on $\mathbb{R}^n$.

$$p_n(f) = \max \left\{ \left| \frac{\partial^{|I|} f}{\partial x^I}(x) \right| : \|x\| \leq n, |I| \leq n \right\}$$

($I = (i, \ldots, i_n)$ and $|I| = i_1 + \cdots + i_n$ comes from standard multiindex notation.) This example can be expanded to a C^{∞} manifold which is σ-compact.

V. $\mathscr{S}(\mathbb{R})$, the Schwartz space of rapidly decreasing functions on $\mathbb{R}$. The nth seminorm is

$$p_n(f) = \sup_{\substack{x \in \mathbb{R} \\ 0 \leq j \leq n}} (1 + |x|)^n |f^{(j)}(x)|.$$

$\mathscr{S}(\mathbb{R})$ is defined as the subset of $C^{\infty}(\mathbb{R})$ for which these seminorms are all finite.

VI. (From Sect. 3.1). Suppose m is Lebesgue measure on $[0, 1]$.

$$X = \bigcap_{p>1} L^p(m).$$

The nth seminorm is just $\| \bullet \|_n$. It is left to the reader to check that these norms suffice.

3.8   LF-Spaces

As noted in the Preface, a majority of the topological vector spaces used in analysis are Banach spaces. Also, a majority of the remaining spaces are Fréchet spaces. In fact, nearly all the spaces routinely used in analysis are one of four types: Banach spaces, Fréchet spaces, LF-spaces, or the dual spaces of Fréchet spaces or LF-spaces. (The dual space of a Banach space, of course, is another Banach space.)

There is a conundrum associated with the definition of an LF-space, which also arises (but is resolved differently) in commutative algebra. Nobody would disagree with the idea that every principal ideal domain is a unique factorization domain, but there is a (slight) discomfort in noticing that every field is a principal ideal domain. The discomfort arises from vacuity (e.g., "Every nonzero nonunit is a prime or a product of primes.") or exceptions (e.g., "A principal ideal domain has Krull dimension one—unless it is a field, which has Krull dimension zero"). Here, a Banach space is always a Fréchet space—nobody disagrees with that. However, an incomplete reading of the definition may lead one to conclude that a Fréchet space is automatically an LF-space. In fact, that is excluded. (Note: Some authors do not exclude it, and use the term "strict LF-space" for what we call an LF-space. They also [usually] allow examples such as that of Exercise 18.)

Definition 3.38. An **LF-space** is a vector space X over $\mathbb{R}$ or $\mathbb{C}$ for which

$$X = \bigcup_{n=1}^{\infty} X_n,$$

where each X_n is a subspace of X equipped with a Hausdorff, locally convex topology making each X_n into a Fréchet space, subject to the following three constraints:

1. Each $X_n \subset X_{n+1}$.
2. The topology X_{n+1} induces on X_n is its Fréchet space topology.
3. $X_n \neq X$ for all n.

The **LF-topology** on X is defined via Theorem 3.2, using the base:

$$\mathscr{B}_0 = \{B \subset X : B \text{ is convex and balanced, and } B \bigcap X_n \text{ is a}$$

$$\text{neighborhood of } 0 \text{ in the Fréchet topology of } X_n.\}$$

Observe that since each X_n is complete, it is closed (Proposition 1.30), so thanks to constraint 3, an LF-space is *always* first category. One more definition, which we return to later and in Sect. 4.3: an **LB-space** is an LF-space in which the subspaces X_n are actually Banach spaces.

Examples of LF-Spaces

I. $\mathbb{R}[x]$ or $\mathbb{C}[x]$. (Polynomials.) Here, X_n consists of polynomials of degree $\leq n$. The topology on X_n can be taken as its Euclidean topology, since $X_n \approx \mathbb{F}^{n+1}$:

$$\sum_{j=0}^{n} a_j x^j \leftrightarrow \begin{pmatrix} a_0 \\ a_1 \\ \vdots \\ a_n \end{pmatrix}.$$

II. $C_c(H)$, the continuous functions with compact support on a noncompact, locally compact, σ-compact Hausdorff space. H can be written as

$$H = \bigcup_{n=1}^{\infty} K_n$$

where each K_n is compact, and each $K_n \subset \text{int}(K_{n+1})$. Set

$$X_n = \{f \in C_c(H) : \text{supp}(f) \subset K_n\}.$$

The topology on X_n is given by $\|f\| = \max |f(K_n)|$.

III. $C_c^{\infty}(\mathbb{R}^m)$, the space of C^{∞} functions on $\mathbb{R}^m$ with compact support. Letting B_r^- denote the closed ball of radius r,

$$X_n = \{f \in C_c^{\infty}(\mathbb{R}^m) : \text{supp}(f) \subset B_n^-\}.$$

The Fréchet topology on X_n is the one it gets as a (closed) subspace of $C^{\infty}(\mathbb{R}^m)$. This example can be expanded to a C^{∞} manifold that is σ-compact but not compact.

Examples I and II above are LB-spaces, while Example III is not. Example III is so important that it is conceivable that LF-spaces, as a class of locally convex spaces, would be defined even if Example III were the only example.

One other feature stands out about the examples: The inclusions $X_n \hookrightarrow X_{n+1}$ are actually isometries, when one uses the metrics associated with the [semi]norm[s] used in the previous section. It turns out that this can always be arranged (Exercise 14), but this fact is not particularly useful.

It is immediately evident that the topology given in Definition 3.38 works: $\mathscr{B}_0$ is evidently closed under intersections and multiplication by $\frac{1}{2}$. However, anything else will take some work. There is a fundamental construction that needs a lot of discussion. This construction basically works between X_n and X_{n+1}; it will be referred to as the "link" construction (not a standard term), because it will form a link in the chain of containments

$$X_k \subset X_{k+1} \subset X_{k+2} \subset \cdots \subset X_n \;\; \subset \;\; X_{n+1} \subset \cdots$$

$$\underset{\text{here}}{\uparrow}$$

Given: a convex, balanced neighborhood U_n of 0 in the Fréchet space X_n, and another convex, balanced neighborhood U_{n+1} of 0 in the Fréchet space X_{n+1}, subject to the condition: $U_{n+1} \cap X_n \subset U_n$. Set

$$L(U_n, U_{n+1}) = \{tx + (1-t)y : x \in U_n, y \in U_{n+1}, t \in [0,1)\}.$$

The basic properties here are easily established, and will be immediately used in the next proposition. These results will subsequently be referred to by their "Fact number."

1. $L(U_n, U_{n+1})$ *is convex and balanced, and* $L(U_n, U_{n+1}) \supset U_{n+1}$.
   $L(U_n, U_{n+1})$ is convex by Proposition 2.14, while if $x \in U_n$, $y \in U_{n+1}$, $t \in [0,1)$, and $|c| \le 1$, then

   $$c(tx + (1-t)y) = t(cx) + (1-t)(cy) \in L(U_n, U_{n+1}).$$

   Finally, taking $t = 0$, $y \in L(U_n, U_{n+1})$.

2. $[0,1)U_n$ *is contained in the interior in* X_{n+1} *of* $L(U_n, U_{n+1})$:
   If $0 \le t < 1$, then for all $x \in U_n$, $tx + (1-t)U_{n+1}$ is a neighborhood of tx in X_{n+1}, and $tx + (1-t)U_{n+1} \subset L(U_n, U_{n+1})$.

3. $L(U_n, U_{n+1}) \cap X_n \subset U_n$:
   If $0 \le t < 1$, $x \in U_n$, and $y \in U_{n+1}$, with $z = tx + (1-t)y \in X_n$, then $y = (1-t)^{-1}(z - tx) \in X_n$, so $y \in U_{n+1} \cap X_n \subset U_n$. Hence $z \in U_n$ since U_n is convex.

4. *If* U_n *is open in* X_n, *then* $L(U_n, U_{n+1}) \cap X_n = U_n$.
   From facts 2 and 3, $[0,1)U_n \subset L(U_n, U_{n+1}) \cap X_n \subset U_n$. But $U_n = [0,1)U_n$ when U_n is open (Theorem 2.15).

5. *If* U_n *is open in* X_n, *then* $L(U_n, U_{n+1}) = \text{con}(U_n \cup U_{n+1})$, *the convex hull of* U_n *and* U_{n+1}:
   $L(U_n, U_{n+1})$ is convex (Fact 1), and everything in $L(U_n, U_{n+1})$ lies in $\text{con}(U_n \cup U_{n+1})$ by Proposition 2.12. On the other hand, $U_n \subset L(U_n, U_{n+1})$ when U_n is open, and $U_{n+1} \subset L(U_n, U_{n+1})$ by Fact 1, so $U_n \cup U_{n+1} \subset L(U_n, U_{n+1})$, giving $\text{con}(U_n \cup U_{n+1}) \subset L(U_n, U_{n+1})$.

6. *If* U_{n+1} *is open in* X_{n+1}, *then* $L(U_n, U_{n+1})$ *is open in* X_{n+1}:

   $$L(U_n, U_{n+1}) = \bigcup_{0 \le t < 1} \bigcup_{x \in U_n} (tx + (1-t)U_{n+1}).$$

It seems clear at this point that the link construction works best with open sets, and for the full chain construction we shall make that restriction. Suppose we have a sequence of convex, balanced sets $\langle U_n \rangle$, starting at some k and going to ∞, where each U_n is open in X_n, with $U_n \supset X_n \cap U_{n+1}$. Recursively define

$$V_k = U_k; V_{n+1} = L(V_n, U_{n+1}).$$

The "chain" is $\langle V_n \rangle$, and will be so referred to in the next few "facts." The next two make our construction legitimate.

7. *For all* $n \geq k$, $U_n \subset V_n$:

   True when $n = k$ by definition; true for $n + 1$ by Fact 1. (This is not an induction.)

8. *For all* $n \geq k$, $U_{n+1} \cap X_n \subset V_n$:

   $U_{n+1} \cap X_n \subset U_n$ by assumption, so that $U_{n+1} \cap X_n \subset U_n \subset V_n$ by Fact 7.

9. *For all* $n \geq k$, V_n *is open in* X_n.

   True when $n = k$ by definition; true for $n + 1$ by Fact 6. (This is not an induction, either.)

10. *For all* $n \geq k$, $V_{n+1} \cap X_n = V_n$:

    True by Fact 4, which applies by Fact 9.

11. *For all* $n \geq m \geq k$, $V_n \cap X_m = V_m$:

    True when $n = m$ by definition; if $V_n \cap X_m = V_m$, then $V_{n+1} \cap X_m = V_{n+1} \cap X_n \cap X_m = V_n \cap X_m = V_m$ by Fact 10. (This *is* an induction, on n.)

12. $V = \bigcup\limits_{n=k}^{\infty} V_n$ *is a convex, balanced, LF-neighborhood of* 0 *in* X, *for which* $V \cap X_m = V_m$ *for each* $m \geq k$:

    This is an ascending (Fact 10) union of convex balanced sets, so it is convex and balanced. Since it is ascending, for any $m \geq k$:

$$V \cap X_m = \left(\bigcup_{n=m}^{\infty} V_n \right) \cap X_m = \bigcup_{n=m}^{\infty} (V_n \cap X_m) = V_m.$$

It follows that V belongs to our LF base.

We are now ready for our first main result.

Proposition 3.39. *Suppose X is an LF-space, constructed from an ascending union of Fréchet subspaces $\langle X_n \rangle$. Then:*

(a) *The LF topology on X induces the (original) Fréchet topology on each X_k.*

(b) *X is Hausdorff.*

(c) *If $Y_1 \subset Y_2 \subset \cdots$ is an ascending sequence of subspaces, with $X = \bigcup Y_n$, and each Y_k being a Fréchet space in the induced topology, then*

   (c1) *$\forall k \, \exists n : X_k \subset Y_n$.*

   (c2) *$\forall n \, \exists k : Y_n \subset X_k$.*

   (c3) *X has the LF-topology associated with the sequence of subspaces $\langle Y_k \rangle$.*

Proof. (a) If $B \in \mathscr{B}_0$, where $\mathscr{B}_0$ is the base defined in Definition 3.38, then $B \cap X_k$ is a neighborhood of 0 in X_k, so the induced topology is at least coarser than the Fréchet topology on X_k. But if U_k is any open, convex, balanced neighborhood of 0 in X_k, one can define a sequence $\langle U_n \rangle$ recursively as follows: U_k is given, and for each n:

U_n *is an open, convex, balanced neighborhood of* 0 *in* X_n, *which has the topology induced from* X_{n+1}, *so one can choose an open, convex, balanced neighborhood* U_{n+1} *of* 0 *in* X_{n+1} *for which* $U_{n+1} \cap X_n \subset U_n$.

Now form the chain construction to $\langle U_n \rangle$, manufacturing $\langle V_n \rangle$. $V = \bigcup V_n \in \mathscr{B}_0$, and $V \cap X_k = V_k = U_k$ by Fact 12. Hence the induced topology is finer than the Fréchet topology on X_k.

(b) Given $x \neq 0$: $x \in X_k$ for some k. Choose an open, convex, balanced neighborhood U_k of 0 in X_k for which $x \notin U_k$, and repeat the construction above for part (a): $x \notin V$, so $x \notin \{0\}^-$ (Proposition 1.9), so X is Hausdorff (Corollary 1.11).

(c1) Each Y_n is complete, hence closed in X (Proposition 1.30), so $Y_n \cap X_k$ is closed in X_k by part (a). Since $X = \bigcup Y_n$,

$$X_k = \left(\bigcup_{n=1}^{\infty} Y_n \right) \cap X_k = \bigcup_{n=1}^{\infty} (Y_n \cap X_k),$$

so $\exists\, n$, for which the interior of $(Y_n \cap X_k)$ in X_k is nonempty (Baire category). That is, $[0, 1)(Y_n \cap X_k)$ is an open neighborhood of 0 in X_k (Theorem 2.15). But $Y_n \cap X_k$ is a subspace, so $[0, 1)(Y_n \cap X_k) = Y_n \cap X_k$, so $Y_n \cap X_k$ is an open subspace of X_k, and so $Y_n \cap X_k = X_k$, giving $X_k \subset Y_n$.

(c2) Each X_k is complete, hence closed in X (Proposition 1.30), so $X_k \cap Y_n$ is closed in Y_n by assumption. Since $X = \bigcup X_k$:

$$Y_n = \left(\bigcup_{k=1}^{\infty} X_k \right) \cap Y_n = \bigcup_{k=1}^{\infty} (X_k \cap Y_n),$$

so $\exists\, k$ for which the interior of $(X_k \cap Y_n)$ in Y_n is nonempty (Baire category). [The rest of the argument is as for (c1), with Y_n and X_k reversed.]

(c3) Set

$$\mathscr{B}_1 = \{B \subset X : B \text{ is convex and balanced, and}$$

$$B \cap Y_n \text{ is a neighborhood of 0 in } Y_n.\}.$$

If $B \in \mathscr{B}_0$, then $\forall\, n$, $\exists\, k$ with $Y_n \subset X_k$ [part (c2)], and $B \cap X_k$ is a neighborhood of 0 in X_k, so $B \cap Y_n = (B \cap X_k) \cap Y_n$ is a neighborhood of 0 in Y_n (Y_n has the topology induced from X_k by part (a) and "transitivity" for induced topologies). Hence $B \in \mathscr{B}_1$.

If $B \in \mathscr{B}_1$, then $\forall\, k$, $\exists\, n$ with $X_k \subset Y_n$ [part (c1)], and $B \cap Y_n$ is a neighborhood of 0 in Y_n, so $B \cap X_k = (B \cap Y_n) \cap X_k$ is a neighborhood of 0 in X_k (transitivity again). Hence $B \in \mathscr{B}_0$.

So: $\mathscr{B}_0 = \mathscr{B}_1$, so the LF topologies coincide. $\qquad\qquad\square$

The proof of part (c3) gives some glimmer of why, in defining $\mathscr{B}_0$, we did *not* assume that the intersection $B \cap X_k$ was an *open* neighborhood of 0 in X_k, hence (eventually) compatible with the chain construction. This glimmer will become even clearer in Sect. 4.1.

In what follows, we shall write "$X = \bigcup X_n$ is an LF-space," meaning $\langle X_n \rangle$ is an ascending sequence of Fréchet spaces for which the union is X, with all the assumptions made for constructing an LF-space. What part (c3) does is make this unambiguous: X is [in some way] an LF-space, and $\langle X_n \rangle$ is an ascending sequence of Fréchet subspaces, with the union X.

Proposition 3.40. *Suppose* $X = \bigcup X_n$ *is an LF-space, and* U *is a convex subset of* X. *Then* U *is open in* X *if, and only if,* $U \cap X_n$ *is open in* X_n *for all* n.

Proof. If U is open in X, then $U \cap X_n$ is open in X_n for all n, by Proposition 3.39(a). On the other hand, suppose U is convex in X, and $U \cap X_n$ is open in X_n for all n. Suppose $x \in U$; then $x \in X_k$ for some k, so $(U - x) \cap X_k$ is an open neighborhood of 0 in X_k. Choose an open, convex balanced neighborhood U_k of 0 in X_k, with $U_k \subset (U - x) \cap X_k = (U \cap X_k) - x$. Now extend this to a sequence $\langle U_n \rangle$ by requiring that U_{n+1} be an open, convex, balanced neighborhood of 0 in X_{n+1} for which

$$U_{n+1} \subset (U - x) \cap X_{n+1} = (U \cap X_{n+1}) - x \text{ and } U_{n+1} \cap X_n \subset U_n.$$

Forming the chain construction to produce $\langle V_n \rangle$ and V, each $V_n \subset U - x$ by induction on n: if $n = k$, then $V_k = U_k \subset U - x$, while if $V_n \subset U - x$, then $V_{n+1} = L(V_n, U_{n+1}) \subset U - x$ by Fact 5, since $U - x$ is convex. Hence $V = \bigcup V_n \subset U - x$, so (since $V \in \mathscr{B}_0$), $x + V \subset U$, so that $x \in \text{int}(U)$. Since x was arbitrary, U consists of interior points, and so is open. □

By the way, the preceding is sometimes used to define the LF topology. It is, however, a bit difficult to work with initially. It is handy for many applications:

Corollary 3.41. *Suppose* $X = \bigcup X_n$ *is an LF-space,* Y *is a locally convex space, and* $T : X \to Y$ *is a linear transformation. Then* T *is continuous if, and only if,* $T|_{X_n}$ *is continuous on* X_n *for each* n.

Proof. If T is continuous on X, then $T|_{X_n}$ is continuous on X_n by definition of the induced topology, so suppose $T|_{X_n}$ is continuous on X_n for all n. If V is an open convex neighborhood of 0 in Y, then $T|_{X_n}^{-1}(V)$ is open in X_n, that is $T^{-1}(V) \cap X_n$ is open in X_n. Since $T^{-1}(V)$ is convex [Proposition 2.17(c)], $T^{-1}(V)$ is open in X by Proposition 3.40, and so T is continuous by Proposition 1.26(a). □

Our next subject is completeness. For this, we will need several more little facts, this time concerning nets. We will also need a lemma about the chain construction. First, the lemma:

Lemma 3.42. *Suppose* $X = \bigcup X_n$ *is an LF-space. Concerning the chain construction, starting at* $k = 1$:

(a) *If* U *is an open, convex, balanced neighborhood of* 0 *in* X, *then setting* $U_n = U \cap X_n$ *and applying the chain construction produces (in the earlier notation)* $V_n = U_n$ *and* $V = U$.

(b) *If* $\langle U_n \rangle$ *and* $\langle \tilde{U}_n \rangle$ *are sequences of convex, balanced, open sets to which the chain construction is applicable, yielding* $\langle V_n \rangle$ *and* $\langle \tilde{V}_n \rangle$, *respectively, then*

$$(\forall n : U_n \subset \tilde{U}_n) \Rightarrow (\forall n : V_n \subset \tilde{V}_n).$$

(c) *If* $\langle U_n \rangle$ *is a sequence of convex, balanced, open sets to which the chain construction is applicable, yielding* $\langle V_n \rangle$ *and* $V = \bigcup V_n$, *then* $X_\ell + V = \tilde{V}$, *where one sets*

$$\tilde{U}_n = \begin{cases} X_n \ if \ n \le \ell \\ U_n \ if \ n > \ell \end{cases}$$

and applies the chain construction to $\langle \tilde{U}_n \rangle$, *yielding* $\langle \tilde{V}_n \rangle$ *and then* $\tilde{V}$.

Proof. (a) $U_n = V_n$ by induction on n. $n = 1 : U_1 = V_1$ by definition. If $U_n = V_n$, then since $U_n = U \cap X_n = U \cap X_{n+1} \cap X_n = U_{n+1} \cap X_n$, $V_{n+1} = L(V_n, U_{n+1}) = \mathrm{con}(U_n \bigcup U_{n+1}) = \mathrm{con}(U_{n+1}) = U_{n+1}$ by Fact 5, since U_{n+1} is convex.

(b) If $U_n \subset \tilde{U}_n$ for all n, then $V_n \subset \tilde{V}_n$ for all n by induction on n. $n = 1 : V_1 = U_1 \subset \tilde{U}_1 = \tilde{V}_1$. If $V_n \subset \tilde{V}_n$, then $V_{n+1} = L(V_n, U_{n+1}) = \mathrm{con}(V_n \bigcup U_{n+1}) \subset \mathrm{con}(\tilde{V}_n \bigcup \tilde{U}_{n+1}) = L(\tilde{V}_n, \tilde{U}_{n+1}) = \tilde{V}_{n+1}$ by Fact 5.

(c) $U^{\#} = X_\ell + V$ is an open, convex, balanced set, so one can set $U_n^{\#} = U^{\#} \cap X_n$ and produce $\langle U_n^{\#} \rangle = \langle V_n^{\#} \rangle$ yielding $V^{\#} = U^{\#}$ by part (a). But for all n, $U_n \subset V_n$ (Fact 7), so $U_n \subset V_n \subset V \subset U^{\#}$, so $U_n \subset U_n^{\#}$. Hence $\tilde{U}_n \subset U_n^{\#}$ for all n, yielding $\tilde{V} \subset V^{\#} = U^{\#}$ by part (b). Hence $\tilde{V} \subset X_\ell + V$.

On the other hand, $U_n \subset \tilde{U}_n$ for all n, yielding $V_n \subset \tilde{V}_n$ for all n and $V \subset \tilde{V}$, by part (b). Also, $X_\ell \subset \tilde{U}_\ell \subset \tilde{V}_\ell \subset \tilde{V}$, so $X_\ell \bigcup V \subset \tilde{V}$. But V is open, so if $x \in V = [0, 1)V$ (Theorem 2.15), writing $x = ty$, $y \in V$, and $t \in [0, 1)$; and if $z \in X_\ell$, then

$$z + x = (1 - t) \cdot \left(\frac{1}{(1-t)} z \right) + ty \in \tilde{V}$$

since $\tilde{V}$ is convex. Hence $X_\ell + V \subset \tilde{V}$. $\square$

Now for the little facts concerning nets. Suppose $\langle x_\alpha \rangle$ is a net in our LF-space X, defined on a directed set D. If $\langle y_\alpha \rangle$ is another net defined on D, declare that $\langle x_\alpha \rangle \sim \langle y_\alpha \rangle$ when $\lim(y_\alpha - x_\alpha) = 0$. It is not hard to see that this is an equivalence relation (Exercise 16), but all we need is symmetry, which really is "obvious."

13. *If $\langle x_\alpha \rangle \sim \langle y_\alpha \rangle$ and $\lim x_\alpha = x$, then $\lim y_\alpha = x$.*

   Note that $\lim(x_\alpha, y_\alpha - x_\alpha) = (x, 0)$ in $X \times X$ by Proposition A.2 in Appendix A, so $\lim y_\alpha = x$ by continuity of addition.

14. *If $\langle x_\alpha \rangle \sim \langle y_\alpha \rangle$ and $\langle x_\alpha \rangle$ is a Cauchy net, then $\langle y_\alpha \rangle$ is a Cauchy net.*

   If U is a convex balanced neighborhood of 0, $\exists \alpha$ s.t. $\beta, \gamma \succ \alpha \Rightarrow x_\beta - x_\gamma \in \frac{1}{3} U$. Also, $\exists \alpha'$ s.t. $\beta \succ \alpha' \Rightarrow y_\beta - x_\beta \in \frac{1}{3} U$, so that also $x_\beta - y_\beta \in \frac{1}{3} U$ since U is balanced. $\exists \alpha_0 \in D$ s.t. $\alpha_0 \succ \alpha$ and $\alpha_0 \succ \alpha'$ since D is directed. If $\beta, \gamma \succ \alpha_0$, then $\beta, \gamma \succ \alpha$ (so that $x_\beta - x_\gamma \in \frac{1}{3} U$) and $\beta, \gamma \succ \alpha'$ (so that $y_\beta - x_\beta \in \frac{1}{3} U$ and $x_\gamma - y_\gamma \in \frac{1}{3} U$) so that

$$y_\beta - y_\gamma = (y_\beta - x_\beta) + (x_\beta - x_\gamma) + (x_\gamma - y_\gamma) \in \frac{1}{3} U + \frac{1}{3} U + \frac{1}{3} U = U$$

by Proposition 3.3(b).

   Finally, suppose D' is another nonempty directed set, for which the members are denoted by capital letters for reasons that will become clear. $D \times D'$ is now directed, where $(\alpha, U) \succ (\beta, V)$ when $\alpha \succ \beta$ *and* $U \succ V$: given (α, U) and (β, V), $\exists \gamma \in D$ s.t. $\gamma \succ \alpha$ and $\gamma \succ \beta$ since D is directed, and $\exists W \in D'$ s.t. $W \succ U$ and $W \succ V$ since D' is directed, so that $(\gamma, W) \succ (\alpha, U)$ and $(\gamma, W) \succ (\beta, V)$. If $\langle x_\alpha \rangle$ is a net defined on D, one can define $\langle x_\alpha \rangle$ on $D \times D'$ by setting $x_{\alpha, V} = x_\alpha$; this net will be denoted by $\langle x_\alpha : D \times D' \rangle$, and the limit by $\lim_{D \times D'} x_\alpha$.

15. *If $\langle x_\alpha : \alpha \in D \rangle$ is a net, and D' is a directed set, and $x \in X$, then*

$$\lim_D x_\alpha = x \Leftrightarrow \lim_{D \times D'} x_\alpha = x.$$

   Pick any $U \in D'$. (Remember, directed sets are nonempty.) If N is a neighborhood of x, then $(\beta \succ \alpha \Rightarrow x_\beta \in N) \Leftrightarrow ((\beta, V) \succ (\alpha, U) \Rightarrow x_\beta \in N)$, since x_β does not depend on V.

16. *If $\langle x_\alpha : \alpha \in D \rangle$ is a net, and D' is a directed set, then $\langle x_\alpha : \alpha \in D \rangle$ is a Cauchy net $\Leftrightarrow \langle x_\alpha : D \times D' \rangle$ is a Cauchy net.*

   Pick any $U \in D'$. If N is a neighborhood of 0, then $(\beta, \gamma \succ \alpha \Rightarrow x_\beta - x_\gamma \in N) \Leftrightarrow ((\beta, V), (\gamma, W) \succ (\alpha, U) \Rightarrow x_\beta - x_\gamma \in N)$, since x_α and x_γ do not depend on V and W.

Theorem 3.43. *LF-spaces are complete.*

Proof. Suppose $\langle x_\alpha \rangle$ is a Cauchy net in an LF-space $X = \bigcup X_n$, defined on a directed set D. Let D' be a neighborhood base at 0 in X consisting of open, convex, balanced sets. If $U, V \in D'$, declare $U \succ V$ when $U \subset V$ (directed downward). D' is now a nonempty directed set. If $\alpha \in D$ and $U \in D'$, let $n(\alpha, U)$ denote the smallest n for which $X_n \cap (x_\alpha + U) \neq \emptyset$. There are such n's; the n for which $x_\alpha \in X_n$ is one. Choose

$$y_{\alpha, U} \in X_{n(\alpha, U)} \cap (x_\alpha + U).$$

$\langle y_{\alpha,U} \rangle$ is now a net defined on $D \times D'$. Furthermore, $y_{\alpha,U} \in x_\alpha + U \Rightarrow y_{\alpha,U} - x_\alpha \in U$, so almost trivially,

$$\lim_{D \times D'} (y_{\alpha,U} - x_\alpha) = 0.$$

So: $\langle y_{\alpha,U} \rangle \sim \langle x_\alpha : D \times D' \rangle$. Thus $\langle y_{\alpha,U} \rangle$ is a Cauchy net by Facts 16 and 14.

Set $(D \times D')_m = \{(\alpha, U) \in D \times D' : n(\alpha, U) \leq m\}$. If some $(D \times D')_m$ is cofinal in $D \times D'$, then we are done: $\langle y_{\alpha,U} \rangle$ is now a Cauchy net in X_m on $(D \times D')_m$, since $(\alpha, U) \in (D \times D')_m \Rightarrow y_{\alpha,U} \in X_{n(\alpha,U)} \subset X_m$. But X_m is complete, as are all Fréchet spaces, so $\exists x \in X_m$ for which

$$\lim_{(D \times D')_m} y_{\alpha,U} = x \qquad \text{(limit in } X_m\text{)},$$

$$\text{giving } \lim_{(D \times D')_m} y_{\alpha,U} = x \qquad \text{(limit in } X\text{)},$$

since X_m has the induced topology [Proposition 3.39(a)]. But now $\lim_{D \times D'} y_{\alpha,U} = x$ by Corollary 1.33. Hence $\lim_{D \times D'} x_\alpha = x$ by Fact 13, and $\lim_D x_\alpha = x$ by Fact 15.

The final step is to show that, in fact, some $(D \times D')_m$ must be cofinal. So suppose not; suppose no $(D \times D')_m$ is cofinal. Define a sequence (α_n, W_n) as follows, constructed so that $(\alpha_1, W_1) \prec (\alpha_2, W_2) \prec \cdots$:

$(D \times D')_1$ is not cofinal, so $\exists (\alpha_1, W_1)$ s.t. $(\alpha_1, W_1) \not\prec (\beta, V)$ for all $(\beta, V) \in (D \times D')_1$. Given $(\alpha_1, W_1) \prec (\alpha_2, W_2) \prec \cdots \prec (\alpha_n, W_n)$: $(D \times D')_{n+1}$ is not cofinal, so $\exists (\beta, V) \in D \times D'$ such that $(\gamma, W) \succ (\beta, V) \Rightarrow (\gamma, W) \notin (D \times D')_{n+1}$. Choose such a (β, V), and choose (α_{n+1}, W_{n+1}) so that $(\alpha_{n+1}, W_{n+1}) \succ (\beta, V)$ and $(\alpha_{n+1}, W_{n+1}) \succ (\alpha_n, W_n)$. Note that in all cases, $(\gamma, W) \succ (\alpha_n, W_n) \Rightarrow (\gamma, W) \notin (D \times D')_n$.

Now for the trick. $(\alpha_1, W_1) \prec (\alpha_2, W_2) \prec \cdots$, so $W_1 \supset W_2 \supset \cdots$. Apply the chain construction to the sequence $U_n = W_n \cap X_n$: D' consists of open, convex, balanced sets, so each $W_n \cap X_n$ is open, convex, and balanced in X_n. Furthermore, $W_n \cap X_n \supset W_{n+1} \cap X_n = (W_{n+1} \cap X_{n+1}) \cap X_n$, so these sets are suitable for applying the chain construction, producing $\langle V_n \rangle$, and V. This V is a perfectly legitimate neighborhood of 0 in X, and our original $\langle x_\alpha \rangle$ was a Cauchy net, so $\exists \alpha$ s.t $\beta, \gamma \succ \alpha \Rightarrow (x_\beta - x_\gamma) \in V$. This x_α is in X, so it lies in some X_ℓ. Now $\exists \beta \in D$ s.t. $\beta \succ \alpha$ and $\beta \succ \alpha_\ell$ (same ℓ) since D is directed. The contradiction will come from where x_β must lie.

First, $\beta \succ \alpha$ and $\alpha \succ \alpha$, so $x_\beta - x_\alpha \in V$. Now $x_\alpha \in X_\ell$, so $x_\beta \in x_\alpha + V \subset X_\ell + V$.

Next, $X_\ell + V$ is obtained, by Lemma 3.42(c), as the $\tilde{V} = \bigcup \tilde{V}_n$, using the sets

$$\tilde{U}_n = \left\{ \begin{array}{ll} X_n & \text{if } n \leq \ell \\ W_n \cap X_n & \text{if } n > \ell \end{array} \right\}$$

while W_ℓ and $X_\ell + W_\ell$ are obtained using the sequences $\langle W_\ell \cap X_n \rangle$ and

$$U_n^\# = \begin{cases} X_n & \text{if } n \leq \ell \\ W_\ell \cap X_n & \text{if } n > \ell \end{cases}$$

by Lemma 3.42(a) and (c). But $\tilde{U}_n \subset U_n^\#$ above, so $X_\ell + V \subset X_\ell + W_\ell$.

Finally, $\beta \succ \alpha_\ell$, so $(\beta, W_\ell) \succ (\alpha_\ell, W_\ell)$, so $(\beta, W_\ell) \notin (D \times D')_\ell$, that is $n(\beta, W_\ell) > \ell$. Hence $X_\ell \cap (x_\beta + W_\ell) = \emptyset$, that is $x_\beta \notin X_\ell - W_\ell = X_\ell + W_\ell$ since W_ℓ is balanced.

Where are we? Reordering:

First: $x_\beta \in X_\ell + V$.
Third: $x_\beta \notin X_\ell + W_\ell$.
Second: $X_\ell + V \subset X_\ell + W_\ell$.

Oops!   □

Corollary 3.44. *Suppose $X = \bigcup X_n$ is an LF-space, and A is a bounded set in X. Then $\exists n$ such that $A \subset X_n$.*

Proof. Replace A with $(A^\circ)_\circ$, a nonempty, closed, convex, balanced, bounded set (Proposition 3.19 and Theorem 3.20). This new A is now complete by Theorem 3.43 and Proposition 1.30, so (X_A, p_A) is a Banach space by Proposition 3.30. But $\forall n : X_n \cap X_A$ is p_A-closed in X_A (also by Proposition 3.30), so since $X_A = \bigcup (X_n \cap X_A)$, some $X_n \cap X_A$ has nonempty interior in X_A (Baire category), whence $[0, 1)(X_n \cap X_A) = X_n \cap X_A$ is open in X_A by Theorem 2.15. An open subspace .... Again, $X_n \cap X_A = X_A$, so that $A \subset X_A \subset X_n$.   □

We close with a couple of results about LB-spaces. Suppose $X = \bigcup X_n$ is an LB-space, that is an LF-space in which each X_n is actually a Banach space. Let p_n denote the norm on X_n. Then the topology induced on X_n by X_{n+1} is its original topology. Letting B_n denote the open unit ball in (X_n, p_n), this all means there exist r_n, R_n such that

$$r_n(B_{n+1} \cap X_n) \subset B_n \subset R_n B_{n+1}.$$

Without loss of generality, we can assume that $0 < r_n < 1 < R_n$, and we do assume this.

The sets $B_1, r_1 B_2, r_1 r_2 B_3, \ldots$ are suitable for the chain construction, yielding a sequence $\langle V_n \rangle$ and a set $V = \bigcup V_n$. Now for all $n : r_1 r_2 \cdots r_{n-1} B_n \subset V_n$ ($n = 1\checkmark$; $n > 1$ from Fact 1). But this means that for all $x \in X_n$,

$$\forall t > 0 : x \in t r_1 r_2 \cdots r_{n-1} B_n \Rightarrow x \in t V_n, \text{ or}$$

$$\{t > 0 : x \in t r_1 r_2 \cdots r_{n-1} B_n\} \subset \{t > 0 : x \in t V_n\}, \text{ so that}$$

$$p_{r_1 r_2 \cdots r_{n_1} B_n} \geq p_{V_n}.$$

But for all $r > 0$ and $x \in X_n$:

$$\{t > 0 : tx \in rB_n\} = \{t > 0 : tr^{-1}x \in B_n\}, \text{ so}$$

$$p_{rB_n}(x) = p_{B_n}(r^{-1}x) = r^{-1}p_n(x)$$

since $p_n = p_{B_n}$ is a norm. Hence

$$(r_1 r_2 \cdots r_{n-1})^{-1} p_n \geq p_{V_n}, \text{ or}$$

$$p_n \geq r_1 r_2 \cdots r_{n-1} p_{V_n}.$$

But $V \cap X_n = V_n$ (Fact 12), so $p_V|_{X_n} = p_{V_n}$. We now have that

$$p_n(x) \geq r_1 r_2 \cdots r_{n-1} p_V(x) \text{ for } x \in X_n.$$

But $B_1 \subset R_1 B_2 \subset R_1 R_2 B_3 \subset \cdots \subset R_1 R_2 \cdots R_{n-1} B_n$, and $r_1 B_2 \subset B_2 \subset R_1 B_2 \subset R_1 R_2 B_3 \subset \cdots \subset R_1 R_2 \cdots R_{n-1} B_n$; in general, each $r_1 r_2 \cdots r_{k-1} B_k \subset R_1 R_2 \cdots R_{n-1} B_n$ for $k \leq n$. (This is why we assumed that $r_j < 1 < R_j$.) Hence each $V_k \subset R_1 R_2 \cdots R_{n-1} B_n$ since $R_1 R_2 \cdots R_{n-1} B_n$ is convex, using Fact 5 and induction on k. In particular, $V_n \subset R_1 R_2 \cdots R_{n-1} B_n$, so as above, for $x \in X_n$:

$$p_{V_n}(x) \geq p_{R_1 R_2 \cdots R_{n-1} B_n}(x) = (R_1 R_2 \cdots R_{n-1})^{-1} p_{B_n}(x), \text{ or}$$

$$R_1 R_2 \cdots R_{n-1} p_V(x) \geq p_n(x).$$

What all this means is that $p_V|_{X_n}$ is norm-equivalent to the norm p_n on X_n. We have proved:

Proposition 3.45. *Suppose $X = \bigcup X_n$ is an LB-space. Then there is a single norm $\| \bullet \|$ on X which, when restricted to each X_n, gives its Banach space structure. That is, one may assume without loss of generality that each inclusion $X_n \hookrightarrow X_{n+1}$ is an isometry.*

Warning: *X definitely does **not** have the norm topology. (It is complete and first category.)*

The final result would be slightly easier to prove using material from the next chapter, but it is not difficult here. Consider the result transitional.

Proposition 3.46. *The strong dual of an LB-space is a Fréchet space.*

Proof. Suppose $X = \bigcup X_n$ is an LB-space. Assume we have one norm $\| \bullet \|$ defining the various Banach space structures on the spaces X_n (Proposition 3.45). Let B_n denote the unit ball in X_n. Set

$$\mathscr{B} = \{2^{-k}(B_n)^\circ : n, k \in \mathbb{N}\}.$$

This $\mathcal{B}$ is countable, closed under multiplication by $\frac{1}{2}$, and

$$2^{-j}(B_m)^\circ \bigcap 2^{-k}(B_n)^\circ \supset 2^{-\max(j,k)}(B_{\max(m,n)})^\circ,$$

so $\mathcal{B}$ defines a locally convex topology on X^*. If A is bounded in X, then $A \subset X_n$ for some n (Corollary 3.44), so $A \subset 2^k B_n$ for some k, giving $A^\circ \supset 2^{-k}(B_n)^\circ$. That is, each strong neighborhood of 0 contains a member of $\mathcal{B}$. But each member of $\mathcal{B}$ *is* a strong neighborhood of 0, so $\mathcal{B}$ is a base at 0 for the strong topology, and X^* is first countable. Given the definitions, it remains to show that X^* is complete. Sequentially complete will do, by Theorem 1.34.

Suppose $\langle f_n \rangle$ is a Cauchy sequence in X^*. Then $m, n > N \Rightarrow f_m - f_n \in 2^{-j}(B_k)^\circ$ means $|f_m(x) - f_n(x)| \leq 2^{-j}$ for $x \in B_k$, that is

$$\left\| f_m \big|_{X_k} - f_n \big|_{X_k} \right\|_{op} \leq 2^{-j}.$$

Hence $\langle f_m \big|_{X_k} \rangle$ is a Cauchy sequence in X_k^*, so it converges uniformly on bounded sets in X_k. Letting k float, this gives a pointwise limit f on all of X; moreover, $f \big|_{X_k}$ is the operator norm limit in X_k^*, so $f \big|_{X_k}$ is continuous. Hence f is continuous by Corollary 3.41.   □

In Sect. 4.1, it will be seen that any LF-space satisfies a condition which, in Sect. 4.3, will force its dual space to be complete. We will get to that shortly.

Exercises

1. Let X denote the Banach space of bounded real sequences on $\mathbb{N}$, that is $X = \ell^\infty$. Set $p(\langle x_n \rangle) = \limsup x_n$. Show that p is a gauge.
2. Suppose X is a normed space, with norm $\| \bullet \|$ and open unit ball B. Suppose C is convex, having 0 as an internal point.

   (a) Show that $p_C(x) \leq \|x\|$ for all x if and only if $B \subset C$.
   (b) Show that $p_C(x) \geq \|x\|$ for all x if and only if $C \subset B^-$.

3. Suppose X is a locally convex space, and B is a nonempty closed, convex, balanced subset of X. Suppose D is weak-* dense in B°. Show that $D_\circ = B$.
4. Suppose X is a locally convex space. Show that X^* is weak-′ dense in X'.
5. Suppose X is a locally convex space. Let X_w denote X with the weak topology. Show that $(X_w)^* = X^*$ *without* using Lemma 3.22, but instead by giving two elementary arguments showing that $(X_w)^* \subset X^*$ and $(X_w)^* \supset X^*$.
6. Suppose X is an infinite-dimensional Hausdorff locally convex space.

   (a) Show that the weak-* topology on X^* is not given by a norm.
   (b) Show that if the algebraic dimension of X is uncountable, then the weak-* topology on X^* is not first countable, and hence is not metrizable.

7. Suppose X is a locally convex space, and $\mathscr{T}^*$ is a locally convex topology on X^* which is locally weak-*-closed. Suppose U is a $\mathscr{T}^*$-neighborhood of 0, so that there exists a weak-* closed, $\mathscr{T}^*$-neighborhood V of 0 such that $V \subset U$. Show that there is a strong neighborhood D of 0 such that $D \subset V$. Use this to show that the strong topology on X^* is finer than $\mathscr{T}^*$. *Hint:* V contains a convex, balanced $\mathscr{T}^*$-neighborhood W of 0. Look at $W_\circ$. It may help to show that the set of all weak-* closed, convex, balanced, absorbent sets in X^* constitutes a base at 0 for the strong topology.

8. Suppose X is a locally convex space over $\mathbb{F} = \mathbb{R}$ or $\mathbb{C}$. $X^* = \mathscr{L}_c(X, \mathbb{F})$, of course. Show that the strong topology on X^* is the topology of bounded convergence on $\mathscr{L}_c(X, \mathbb{F})$.

9. Suppose X is a vector space over $\mathbb{R}$ or $\mathbb{C}$, and p is a seminorm on X.

   (a) Set $\ker(p) = \{x \in X : p(x) = 0\}$. Show that $\ker(p)$ is a subspace of X.
   (b) Show that p induces a norm on $X/\ker(p)$, and show that the associated norm topology on $X/\ker(p)$ is the quotient topology.
   (c) Suppose Y is a locally convex space, and $T : Y \to X$ is a linear map. Show that T is continuous if and only if $\pi \circ T$ is continuous, where $\pi : X \to X/\ker(p)$ is the natural projection.

10. Suppose $\mathscr{F}_0$ and $\mathscr{F}_0'$ are two families of seminorms on a vector space X over $\mathbb{R}$ or $\mathbb{C}$, and suppose that $\mathscr{F}_0 \subset \mathscr{F}_0'$. Let $\mathscr{T}$ and $\mathscr{T}'$ denote the locally convex topologies on X produced by $\mathscr{F}_0$ and $\mathscr{F}_0'$ using Definition 3.32 and the discussion following Corollary 3.34. Show that $\mathscr{T} \subset \mathscr{T}'$, with equality if and only if every member of $\mathscr{F}_0'$ is $\mathscr{T}$-continuous.

11. Suppose m is Lebesgue measure on $[0, 1]$. Suppose $1 \leq p < q \leq \infty$. Let A denote the closed unit ball in $L^q(m)$. Show that A is closed in $L^p(m)$, in which space $(L^p(m))_A = L^q(m)$ and $p_A([f]) = \|f\|_q$. (Notation is from Proposition 3.30.)

12. Suppose X is a locally convex space, and p is a seminorm on X. Set $A = \{x \in X : p(x) < 1\}$. Show that $p = p_A$.

13. Suppose X is a locally convex space, and Y is a subspace. Suppose p is a continuous seminorm on Y. Show that p extends continuously to X. *Hint:* Look at the link construction in Sect. 3.8. Use Exercise 12.

14. Suppose $X = \bigcup X_n$ is an LF-space. Show that $\exists$ a countable sequence of seminorms *on* X, $p_1 \leq p_2 \leq \cdots$ such that, for all n, $\{p_1|_{X_n}, p_2|_{X_n}, \ldots\}$ give the Fréchet topology on X_n. *Hint:* Use Exercises 10, 12, and 13, not necessarily in that order.

15. Suppose $X = \bigcup X_n$ is an LF-space and $A \subset X$. Show that A is bounded in X if, and only if, $\exists n$ such that $A \subset X_n$ *and* A is bounded as a subset of the Fréchet space X_n.

16. (See the discussion preceding Theorem 3.43). Suppose G is a topological group with identity 1, and D is a directed set. If $\langle x_\alpha \rangle$ and $\langle y_\alpha \rangle$ are nets defined on D, declare that $\langle x_\alpha \rangle \sim \langle y_\alpha \rangle$ if $\lim x_\alpha^{-1} y_\alpha = 1$. Show that $\sim$ is an equivalence relation.

17. Suppose X is a locally convex space, and B is a neighborhood of 0 in X. Suppose D is a countable dense subset of B. If $x \in D$ and I is an open interval with rational endpoints, set $N(x, I) = \{f \in B^\circ : f(x) \in I\}$.

    (a) Show that the set of all such $N(x, I)$ forms a (countable) subbase for a Hausdorff topology on B° which coincides with the weak-* topology on B°. *Hint:* First show that this topology is coarser than the weak-* topology, then use the Banach–Alaoglu theorem.

    (b) Show that B°, with the weak-* topology, is metrizable. (See Appendix A, Corollary A.5, for the Urysohn metrization theorem.)

18. (This uses some ordinal arithmetic.) Let ω_1 denote the smallest uncountable ordinal, and let μ denote counting measure on $\mathscr{P}(\omega_1)$. Set $X = L^2(\mu)$. If $\alpha < \omega_1$, set $X_\alpha = \{f \in X : f(\beta) = 0 \text{ if } \beta \geq \alpha\}$.

    (a) Show that $X = \bigcup X_\alpha$.

    (b) By analogy with the definition of an LF-space, set $\mathscr{B}_0 = \{B \subset X : B \text{ is convex, and } B \cap X_\alpha \text{ is a neighborhood of 0 in } X_\alpha\}$. Show that $\mathscr{B}_0$ defines a locally convex topology on X.

    (c) Suppose $f : \omega_1 \to (0, 1]$ is a nonincreasing function. Show that f is eventually constant.

    (d) Using (c), show that the topology produced in part (b) is the L^2-topology on X that you started with (!). Suggestion: given $B \in \mathscr{B}_0$, define $f(\alpha)$ to be the largest $r \leq 1$ such that the open ball of radius r in X_α is contained in $B \cap X_\alpha$. (Why *is* there a largest such r?)

19. Suppose X is a Hausdorff locally convex space, and suppose $\langle x_\alpha \rangle$ is a Cauchy net in X that has a weak cluster point x. Show that $\lim x_\alpha = x$ (original topology).
    Hint: Examine the completeness proof in Proposition 3.30, and use the fact that the original topology is locally weakly closed.

20. (Mazur) Suppose X is a metrizable locally convex space, $\langle x_n \rangle$ is a sequence in X, and x is a cluster point of $\langle x_n \rangle$ in the weak topology. Show that there exists a sequence $\langle y_n \rangle$ such that $y_n \to x$ in the original topology, and each

$$y_n \in \operatorname{con}\{x_n, x_{n+1}, x_{n+2}, \ldots\}.$$

Hint: Apply Theorem 3.29 to the original closure of $\operatorname{con}\{x_n, x_{n+1}, x_{n+2}, \ldots\}$.

21. Suppose $(X, \mathscr{B}, \mu)$ is a measure space, $1 \leq p < \infty$, $\langle f_n \rangle$ is a sequence in $L^p(\mu)$, and f is a cluster point of $\langle f_n \rangle$ in the weak topology. Finally, suppose $f_n \to g$ a.e. Show that $f = g$ a.e. *Hint:* Use Exercise 20.

22. Suppose $(X, \mathscr{B}, \mu)$ is a measure space, $1 < p < \infty$, and $\langle f_n \rangle$ is a sequence in $L^p(\mu)$ such that $\|f_n\|_p \leq M$ for all n. ($M = $ a fixed constant.) Finally, suppose $f_n \to g$ a.e. Show that $g \in L^p(\mu)$, $\|g\|_p \leq M$, and $f_n \to g$ weakly. *Hint:* Use Exercise 21, the Banach–Alaoglu theorem, and Proposition A.6 from Appendix A.

Remark: The fact that $\|g\|_p \leq M$ can be derived using integration theory alone. In fact, the whole problem can be done using integration theory, although if μ is infinite, the argument is rather messy.

23. In a sense, the proof of Corollary 3.44 was much more "high powered" than it needed to be: prove Corollary 3.44 without using either the completeness of LF-spaces or the X_A construction by using the following device: if A is not contained in any X_n, then choose $f_n \in X^*$ for which $f_n \big|_{X_n} \equiv 0$ but $f_n(x_n) = n$ for some $x_n \in A$. Look at Σf_n.

24. (Old business) Suppose X is a first countable Hausdorff locally convex space. Show that X has a countable neighborhood base $\mathscr{B} = \{B_1, B_2, B_3, \ldots\}$ at 0 where each B_j is convex and balanced, and $2B_{j+1} \subset B_j$ for each j. Furthermore, show that all the sets B_j may be assumed to be closed, or all the sets B_j may be assumed to be open. (Use Theorem 1.13 plus Proposition 3.1, or Corollary 3.34, your choice.)

Chapter 4
The Classics

4.1 Three Special Properties

There are, of course, a large number of properties that a topological vector space
may have. One has been assumed since the last chapter started: local convexity.
Another will be assumed (mostly) from here on: the Hausdorff condition. There
are plenty of others, but three stand out for their utility concerning the basic (well,
nearly basic) properties of Hausdorff locally convex spaces. In particular, they apply
directly to the "classic" theorems that are the subject of this chapter.

All three have the following form: Any convex balanced set that has the property
that [insert special conditions] is a neighborhood of 0. Before stating the conditions,
recall that a convex balanced set A *absorbs* a set B if $B \subset cA$ for some scalar c. A is
absorbent if it absorbs points. As before, since A is balanced: $B \subset cA \Rightarrow B \subset dA$
whenever $|d| \geq |c|$.

Definition 4.1. Suppose X is a Hausdorff locally convex space.

(i) A **barrel** in X is a subset that is closed, convex, balanced, and absorbent.
(ii) X is **barreled** if each barrel in X is a neighborhood of 0.
(iii) X is **infrabarreled** if each barrel in X that absorbs all bounded sets is a
neighborhood of 0.
(iv) X is **bornological** if each convex, balanced set in X that absorbs all bounded
sets is a neighborhood of 0.

Clearly:

$$\text{barreled} \implies \text{infrabarreled} \impliedby \text{bornological}$$

"Barrelled" is the most important concept, although "bornological" gets the
fastest start. To borrow from Aesop, "barreled" is the tortoise and "bornological"
is the hare—and "infrabarreled" is the gopher, popping up from time to time.

M.S. Osborne, *Locally Convex Spaces*, Graduate Texts in Mathematics 269,
DOI 10.1007/978-3-319-02045-7_4, © Springer International Publishing Switzerland 2014

We start with three propositions. The first two are pretty obvious considerations, but the third is subtle.

Proposition 4.2. *Suppose X is a Hausdorff locally convex space, and Y is a closed subspace.*

(a) If X is barreled, then X/Y is barreled.
(b) If X is infrabarreled, then X/Y is infrabarreled.
(c) If X is bornological, then X/Y is bornological.

Proof. As usual, $\pi : X \to X/Y$ is the quotient map. Suppose B is a convex, balanced subset of X/Y; then $\pi^{-1}(B)$ is also convex and balanced. If $c > 0$ and $x + Y \subset cB$, then $x \in \pi^{-1}(cB) = c\pi^{-1}(B)$ since $\pi^{-1}(B)$ is a union of cosets of Y. In particular, if B absorbs all bounded sets, and A is bounded in X, then B absorbs $\pi(A)$, so $\pi^{-1}(B)$ absorbs A. Finally, if B is closed, then $\pi^{-1}(B)$ is closed. So:

(a) If B is a barrel, then so is $\pi^{-1}(B)$.
(b) If B is a barrel that absorbs all bounded sets, then so is $\pi^{-1}(B)$.
(c) If B is convex and balanced and absorbs all bounded sets, then the same holds for $\pi^{-1}(B)$.

The point is that, whichever case we are in, $\pi^{-1}(B)$ is a neighborhood of 0, so $B = \pi(\pi^{-1}(B))$ is a neighborhood of 0 since π is open. [Proposition 1.26(b)].   □

By the way, there exist barreled spaces with subspaces that are not even infrabarreled; see Köthe [22] for a discussion.

Proposition 4.3. *Suppose X and Y are Hausdorff locally convex spaces.*

(a) If X and Y are barreled, then $X \times Y$ is barreled.
(b) If X and Y are infrabarreled, then $X \times Y$ is infrabarreled.
(c) If X and Y are bornological, then $X \times Y$ is bornological.

Proof. First of all, if A is bounded in X, then any neighborhood of 0 in $X \times Y$ of the form $U \times V$ absorbs $A \times \{0\}$ simply because U absorbs A, so $A \times \{0\}$ is bounded. Suppose B is a convex, balanced subset of $X \times Y$. The slice $B \cap (X \times \{0\})$ can be written as $B_X \times \{0\} = B \cap (X \times \{0\})$; similarly, write $\{0\} \times B_Y = B \cap (\{0\} \times Y)$. Since the projections are homeomorphisms on slices, B closed $\Rightarrow$ B_X and B_Y closed. Finally, B absorbs $A \times \{0\}$ if and only if B_X absorbs A. So:

(a) If B is a barrel, then so are B_X and B_Y.
(b) If B is a barrel that absorbs all bounded sets, then so are B_X and B_Y.
(c) If B is a convex, balanced set that absorbs all bounded sets, then so are B_X and B_Y.

The point is that, whichever case we are in, B_X and B_Y are neighborhoods of 0. But if $x \in B_X$ and $y \in B_Y$, then $\left(\frac{1}{2}x, \frac{1}{2}y\right) = \frac{1}{2}(x, 0) + \frac{1}{2}(0, y) \in B$ since B is convex, so $B \supset \frac{1}{2}(B_X \times B_Y)$, a neighborhood of 0.   □

By the way, this result generalizes considerably, but that requires material from the next section and from the next chapter.

Now for the subtle point. When working over $\mathbb{C}$, the "balanced" condition becomes considerably more restrictive. In particular, the sets required to be neighborhoods of 0 in Definition 4.1 expand when one considers a $\mathbb{C}$-vector space to be an $\mathbb{R}$-vector space. However, the properties are not, in fact, lost.

Proposition 4.4. *Suppose X is a Hausdorff locally convex space over $\mathbb{C}$. Let $X\big|_{\mathbb{R}}$ denote X considered as a locally convex space over $\mathbb{R}$.*

(a) If X is barreled, then $X\big|_{\mathbb{R}}$ is barreled.
(b) If X is infrabarreled, then $X\big|_{\mathbb{R}}$ is infrabarreled.
(c) If X is bornological, then $X\big|_{\mathbb{R}}$ is bornological.

Proof. The fundamentals here appear in Proposition 2.19. Suppose B is a convex, $\mathbb{R}$-balanced subset of X, that is, a convex, balanced subset of $X\big|_{\mathbb{R}}$. Form

$$C = \bigcap_{0 \le \theta < 2\pi} e^{i\theta} B.$$

This C is convex and $\mathbb{C}$-balanced; it is also closed if B is closed. Taking $\mathscr{F}$ to be either the class of singletons or the class of bounded sets:

(a) If B is a barrel in $X\big|_{\mathbb{R}}$, then C is a barrel in X.
(b) If B is a barrel in $X\big|_{\mathbb{R}}$ that absorbs all bounded sets, then C is a barrel in X that absorbs all bounded sets.
(c) If B is convex, balanced set in $X\big|_{\mathbb{R}}$ that absorbs all bounded sets, then C is a convex, balanced set in X that absorbs all bounded sets.

The point is that, whichever case we are in, C is a neighborhood of 0. But $C \subset B$.   □

Now for some particular cases.

Theorem 4.5. *Suppose X is a Hausdorff locally convex space of the second category. Then X is barreled. In particular, Fréchet spaces are barreled.*

Proof. Suppose that X is a Hausdorff locally convex space of the second category, and B is a barrel in X. Since $0 < r < s \Rightarrow rB = s \cdot (r/s)B \subset sB$ (B is balanced),

$$X = \bigcup_{n=1}^{\infty} nB$$

since B is absorbent. Hence $\text{int}(nB) \ne \emptyset$ for some n (second category). But $\text{int}(nB) = n \cdot \text{int}(B)$ since multiplication by n is a homeomorphism. Hence B is a neighborhood of 0 (Theorem 2.15).   □

Corollary 4.6. *LF-spaces are barreled.*

Proof. Suppose that $X = \bigcup X_n$ is an LF-space, and B is a barrel in X. Then for all n, $B \cap X_n$ is a barrel in X_n: It is closed by Proposition 3.39(a), and all else is trivial. By the above, $B \cap X_n$ is a neighborhood of 0 in X_n, so B is, by definition, part of the original neighborhood base at 0 for X.   □

A space need not be complete to be barreled; see Exercises 26–30.

Corollary 4.7 (Absorption Principle). *Suppose X is a Hausdorff locally convex space, and suppose A and B are two closed, convex, balanced subsets of X. Assume that A is bounded and sequentially complete. Then, if B absorbs every point in A, then B absorbs A, that is, $A \subset rB$ for some $r > 0$.*

Proof. If $A = \emptyset$, then $A \subset B$ and we are done, so suppose A is nonempty. As in Proposition 3.30, form the space (X_A, p_A). Then (X_A, p_A) is a Banach space since A is sequentially complete, and $B \cap X_A$ is a barrel in X_A. (B absorbs every point of A, so it absorbs every point of every rA, $r \in \mathbb{R}$.) Hence $B \cap X_A$ is a neighborhood of 0 in X_A, so $B \cap X_A$ absorbs A. Hence B absorbs A.   □

Corollary 4.8. *Suppose X is a Hausdorff locally convex space. If X is infrabarreled and sequentially complete, then X is barreled.*

Proof. Suppose X is an infrabarreled Hausdorff locally convex space that is also sequentially complete, and suppose B is a barrel in X and A is a bounded subset of X. Then $(A^\circ)_\circ$ is closed, convex, balanced, and bounded; it is also sequentially complete: If $\langle x_n \rangle$ is a Cauchy sequence in $(A^\circ)_\circ$, then $x_n \to x$ (there exists $x \in X$) since X is sequentially complete, while $x \in (A^\circ)_\circ$ since $(A^\circ)_\circ$ is closed. Thus, B absorbs $(A^\circ)_\circ$, that is, $A \subset (A^\circ)_\circ \subset cB$, for some $c > 0$. Since A was arbitrary, B absorbs all bounded sets, and so is a neighborhood of 0 since X is infrabarreled.   □

Corollary 4.9. *Infrabarreled spaces are Mackey spaces.*

Proof. Suppose X is an infrabarreled Hausdorff locally convex space. If $D = (D_\circ)^\circ$ is weak-* compact in X^*, then as a subset of X^* with the weak-* topology: D is nonempty, bounded, closed, convex, balanced, and complete (Corollary 1.32), hence D is sequentially complete. If A is bounded in X, then A° is a barrel in X^* which must now absorb D by the absorption principle. Hence $D_\circ$ absorbs A [Theorem 3.20(e)]. Since A is arbitrary and X is infrabarreled, $D_\circ$ is an original neighborhood of 0. Since D was arbitrary, the Mackey topology cannot be strictly finer than the original topology.   □

Now for "bornological," a subject that will reappear in Sect. 4.3, then largely disappear. The bornological condition does not require completeness; far from it.

Proposition 4.10. *Suppose X is a Hausdorff locally convex space. If X is first countable, then X is bornological.*

Proof. Suppose X is a first countable Hausdorff locally convex space. Choose a countable neighborhood base $V_1, V_2, \ldots$ at 0 such that $V_1 \supset V_2 \supset V_3 \supset \cdots$ (Theorem 1.13 provides more, but this is all we need.) Suppose B is a convex, balanced subset of X. Rather than assuming that B absorbs all bounded sets and then proving that B is a neighborhood of 0, we shall assume that B is *not* a neighborhood of 0 and construct a bounded (in fact, compact) set that B does not absorb.

Assume B is not a neighborhood of 0. Then for all n, $\frac{1}{n}V_n \not\subset B$, so $V_n \not\subset nB$. Choose $x_n \in V_n - nB$. Since $x_n \in V_n$ and $V_1 \supset V_2 \supset \cdots : x_n \to 0$, so $\{x_n\} \bigcup \{0\}$ is compact, hence bounded. But $x_n \notin nB$ says that B cannot absorb $\{x_n\} \bigcup \{0\}$. $\qquad\square$

Corollary 4.11. *Normed spaces, Fréchet spaces, and LF-spaces are bornological.*

Proof. Normed spaces and Fréchet spaces are first countable. As for LF-spaces, suppose that $X = \bigcup X_n$ is an LF-space, and B is a balanced, convex subset of X that absorbs all bounded sets. If A is bounded in X_n, then each continuous linear functional on X restricts to a continuous linear functional on X_n (Proposition 3.39) and so is bounded on A. Hence A is bounded in X by Corollary 3.31, so B absorbs A. That is, $B \cap X_n$ is a convex balanced subset of X_n that absorbs A. Since A was arbitrary and X_n is bornological, $B \cap X_n$ is a neighborhood of 0 in X_n. Since n is arbitrary, B is, by definition, a member of the original neighborhood base at 0 defining the LF-topology of X. $\qquad\square$

The preceding proof for LF-spaces works because the LF-topology base $\mathscr{B}_0$ was defined *without* assuming "$B \cap X_n$ is open in X_n" in Sect. 3.8.

Now for the "point" of assuming the bornological condition.

Theorem 4.12. *Suppose X and Y are Hausdorff locally convex spaces, and $T : X \to Y$ is a linear transformation. Consider the following three statements:*

(i) T is continuous.
(ii) If $x_n \to 0$ in X, then $T(x_n) \to 0$ in Y.
(iii) T is bounded, that is $T(A)$ is bounded in Y whenever A is bounded in X.

Then (i) $\Rightarrow$ (ii) $\Rightarrow$ (iii) always, and (iii) $\Rightarrow$ (i) if X is bornological.

Proof. (i) $\Rightarrow$ (ii), since continuity $\Rightarrow$ sequential continuity.

Assume (ii): If A is bounded but $T(A)$ is not bounded in Y, choose a neighborhood V of 0 in Y that does not absorb $T(A)$. choose $T(x_n) \in T(A) - nV$, $x_n \in A$. Then $\frac{1}{n}x_n \to 0$ by Proposition 2.7. But $T\left(\frac{1}{n}x_n\right) \notin V$, so $T\left(\frac{1}{n}x_n\right) \not\to 0$, violating (ii).

Assume (iii), with X bornological. Suppose V is a convex, balanced neighborhood of 0 in Y. If A is bounded in X, then V absorbs $T(A)$, that is $T(A) \subset rV$, so $A \subset rT^{-1}(V)$. Letting A vary, $T^{-1}(V)$ is a convex, balanced set that absorbs all bounded sets, so $T^{-1}(V)$ is a neighborhood of 0 in X. Hence T is continuous [Proposition 1.26(a)]. $\qquad\square$

By the way, the implication (iii) $\Rightarrow$ (i) characterizes the bornological condition; see Exercise 18.

The bornological condition will now take a short rest, then reappear in a fundamental way in Sect. 4.3. After Sect. 4.3, it will largely disappear, and the barreled condition will make its long-term importance quite evident.

4.2 Uniform Boundedness

For locally convex spaces, as for Banach spaces, the notion of "uniform bound-edness" concerns itself with sets of continuous linear transformations. As such, the "right" way to look at such sets is to place them in the appropriate space of continuous linear transformations, then formulate the conditions using functional analysis on that space. However, in practice, the conditions one typically verifies concern boundedness of sets in the range space. These are directly related. The next proposition (and the lemma preceding it) connect these ideas. This proposition is *not* obvious, although it is easy to prove.

As in Sect. 3.6, if X and Y are locally convex spaces, and if $A \subset X$ and $U \subset Y$;

$$N(A, U) = \{T \in \mathscr{L}_c(X, Y) : T(A) \subset U\}.$$

As always, $\mathscr{L}_c(X, Y)$ is the space of continuous linear transformations from X to Y.

Lemma 4.13. *Suppose X and Y are locally convex spaces, and c is a nonzero scalar. Then for all $A \subset X$ and $U \subset Y$:*

$$cN(A, U) = N(A, cU) = N(c^{-1}A, U).$$

Proof. $T \in N(A, cU) \quad \Leftrightarrow \quad T(A) \subset cU \Leftrightarrow c^{-1}T(A) \subset U.$ But

$$c^{-1}T(A) \subset U \quad \Leftrightarrow \quad T(c^{-1}A) \subset U \Leftrightarrow T \in N(c^{-1}A, U) \qquad \text{and}$$

$$c^{-1}T(A) \subset U \quad \Leftrightarrow \quad c^{-1}T \in N(A, U) \Leftrightarrow T \in cN(A, U). \qquad \square$$

Proposition 4.14. *Suppose X and Y are locally convex spaces, and $\mathscr{F} \subset \mathscr{L}_c(X, Y)$.*

(a) *$\mathscr{F}$ is bounded in the topology of pointwise convergence if, and only if, for all $x \in X$ the set $\{T(x) : T \in \mathscr{F}\}$ is bounded in Y.*

(b) *$\mathscr{F}$ is bounded in the topology of bounded convergence if, and only if, for all bounded sets $A \subset X$ the set*

$$\bigcup_{T \in \mathscr{F}} T(A) \text{ is bounded in } Y.$$

Proof. The underlying idea is the same for both parts:

$$\mathscr{F} \subset cN(A, U) \Leftrightarrow \mathscr{F} \subset N(A, cU)$$

$$\Leftrightarrow \forall T \in \mathscr{F} : T(A) \subset cU \Leftrightarrow \bigcup_{T \in \mathscr{F}} T(A) \subset cU.$$

Part (b) is now direct:

$\mathscr{F}$ is bounded in the topology of bounded convergence.

$\Leftrightarrow$ for all bounded $A \subset X$ and convex balanced neighborhood U of 0 in Y: there exists $c > 0$ s.t. $\mathscr{F} \subset cN(A, U)$

$\Leftrightarrow$ for all bounded $A \subset X$ and convex balanced neighborhood U of 0 in Y: there exists $c > 0$ s.t. $\bigcup_{T \in \mathscr{F}} T(A) \subset cU$

$\Leftrightarrow$ for all bounded $A \subset X$: $\bigcup_{T \in \mathscr{F}} T(A)$ is bounded in Y.

As for part (a), the same argument shows that $\mathscr{F}$ is bounded for the topology of pointwise convergence if and only if for all finite $A \subset X$:

$$\bigcup_{T \in \mathscr{F}} T(A) \text{ is bounded in } Y.$$

But

$$\bigcup_{T \in \mathscr{F}} T(A) = \{T(x) : T \in \mathscr{F}, x \in A\} = \bigcup_{x \in A} \{T(x) : T \in \mathscr{F}\},$$

which is bounded if and only if each $\{T(x) : T \in \mathscr{F}\}$ is bounded, since A is finite [Proposition 2.7(c)]. $\qquad\square$

Two-thirds of the next definition will be concerned with giving a shorter phrasing for the conditions appearing above. The other condition is new, and is very important in practice.

Definition 4.15. Suppose X and Y are locally convex spaces, and $\mathscr{F} \subset \mathscr{L}_c(X, Y)$.

(i) $\mathscr{F}$ is **equicontinuous** if, for every neighborhood V of 0 in Y, there exists a neighborhood U of 0 in X such that $T(U) \subset V$ for all $T \subset \mathscr{F}$.
(ii) $\mathscr{F}$ is **bounded for bounded convergence** if $\mathscr{F}$ is bounded in the topology of bounded convergence.
(iii) $\mathscr{F}$ is **bounded for pointwise convergence** if $\mathscr{F}$ is bounded in the topology of pointwise convergence.

Of course, Proposition 4.14 gives equivalent conditions for (ii) and (iii), and there is a *nearly* obvious equivalent condition for (i): If V is a neighborhood of 0 in Y, and $U \subset X$, then

$$\forall T \in \mathscr{F} : T(U) \subset V \Leftrightarrow \forall T \in \mathscr{F} : U \subset T^{-1}(V)$$

$$\Leftrightarrow U \subset \bigcap_{T \in \mathscr{F}} T^{-1}(V).$$

Hence

$\mathscr{F}$ is equicontinuous $\Leftrightarrow$ $\bigcap_{T \in \mathscr{F}} T^{-1}(V)$ is a neighborhood of 0 in X

whenever V is a neighborhood of 0 in Y.

This formulation is the one usually verified in practice.

We can now give the main result of this section. It is more general than the classical Banach–Steinhaus theorem, but there seems to be some disagreement in the literature as to exactly what result should be called the "Banach–Steinhaus theorem" and what should be called the "uniform boundedness theorem." We use both, although the phrase "Equicontinuity theorem" is just as descriptive. Technically, the term "uniform boundedness theorem" applies best to part (c), while "Banach–Steinhaus theorem" applies best to part (d).

Theorem 4.16 (Banach–Steinhaus/Uniform Boundedness Theorem). *Suppose X and Y are Hausdorff locally convex spaces, and $\mathscr{F} \subset \mathscr{L}_c(X, Y)$. Consider the following three conditions on $\mathscr{F}$:*

 (i) *$\mathscr{F}$ is equicontinuous.*
 (ii) *$\mathscr{F}$ is bounded for bounded convergence.*
 (iii) *$\mathscr{F}$ is bounded for pointwise convergence.*

Then:

(a) *(i) $\Rightarrow$ (ii) $\Rightarrow$ (iii) always.*
(b) *If X is infrabarreled, then (ii) $\Rightarrow$ (i).*
(c) *If X is sequentially complete, then (iii) $\Rightarrow$ (ii).*
(d) *If X is barreled, then (i)–(iii) are all equivalent.*

Proof. (a) (ii) $\Rightarrow$ (iii) is trivial, since the topology of bounded convergence is finer than the topology of pointwise convergence. To prove that (i) $\Rightarrow$ (ii), suppose $\mathscr{F}$ is equicontinuous, A is bounded in X, and U is an open neighborhood of 0 in Y. Choose a neighborhood V of 0 in X so that $T(V) \subset U$ for all $T \in \mathscr{F}$. Choose $c > 0$ so that $A \subset cV$. Then $c^{-1}A \subset V$, so that $T(c^{-1}A) \subset U$ for all $T \in \mathscr{F}$, that is $\mathscr{F} \subset N(c^{-1}A, U) = cN(A, U)$. That is, $N(A, U)$ absorbs $\mathscr{F}$.

(d) Assume (iii), and suppose $\mathscr{F}$ is bounded for pointwise convergence and X is barreled. Let U be a barrel neighborhood of 0 in Y. Then

$$B = \bigcap_{T \in \mathscr{F}} T^{-1}(U)$$

is closed, convex, and balanced. If $x \in X$, then there is a $c > 0$ so that $\{T(x) : T \in \mathscr{F}\} \subset cU$, that is $x \in T^{-1}(cU) = cT^{-1}(U)$ for all $T \in \mathscr{F}$. Hence $x \in cB$. That is, letting x vary, B is absorbent, and so is a barrel. Since X is barreled, B is a neighborhood of 0.

(b) Assume (ii), and suppose $\mathscr{F}$ is bounded for bounded convergence and X is infrabarreled. Let U be a barrel neighborhood of 0 in Y. Then as above,

$$B = \bigcap_{T \in \mathscr{F}} T^{-1}(U)$$

is a barrel since $\mathscr{F}$ is bounded for pointwise convergence. Since we are assuming X is infrabarreled, as above, we need only show that B absorbs all bounded sets in X.

Suppose A is bounded in X. Then there is a $c > 0$ so that $\mathscr{F} \subset cN(A, U) = N(c^{-1}A, U)$, so

$$\forall T \in \mathscr{F} : T(c^{-1}A) \subset U, \text{ that is } \forall T \in \mathscr{F} : c^{-1}A \subset T^{-1}(U).$$

Hence $c^{-1}A \subset B$, so that $A \subset cB$.

(c) Assume (iii), and suppose $\mathscr{F}$ is bounded for pointwise convergence and X is sequentially complete. Suppose U is a barrel neighborhood of 0 in Y, and as above set

$$B = \bigcap_{T \in \mathscr{F}} T^{-1}(U).$$

As in the proof of part (d), B is a barrel. If A is bounded in X, then B absorbs each point of $(A^\circ)_\circ$, so B absorbs $(A^\circ)_\circ$ by the absorption principle (Corollary 4.7). Hence B absorbs A. But if $A \subset cB$, then $c^{-1}A \subset B$, so reversing the argument above:

$$\forall T \in \mathscr{F} : c^{-1}A \subset T^{-1}(U), \text{ that is } \forall T \in \mathscr{F} : T(c^{-1}A) \subset U.$$

That is, $\mathscr{F} \subset N(c^{-1}A, U) = cN(A, U)$, so (letting A and U vary) each $N(A, U)$ absorbs $\mathscr{F}$. Hence $\mathscr{F}$ is bounded for bounded convergence.

$\square$

The Banach–Steinhaus theorem has many applications, but one stands out so much that nearly every book on functional analysis includes it: bilinear forms. If X, Y, and Z are locally convex spaces, then a map $f : X \times Y \to Z$ is called *bilinear* if: for all $x \in X$, $f(x, ?) : Y \to Z$ is linear and for all $y \in Y$, $f(?, y) : X \to Z$ is linear. Typically, these two maps are easily checked to be continuous (i.e., f is separately continuous), but what one really wants is joint continuity, or at least something approaching joint continuity.

Proposition 4.17. *Suppose X, Y, and Z are Hausdorff locally convex spaces, and $f : X \times Y \to Z$ is a separately continuous bilinear map. Suppose A is bounded in X, and Y is barreled. Then f is jointly continuous on $A \times Y$.*

Proof. Set $\mathscr{F} = \{f(a, ?) : a \in A\} \subset \mathscr{L}_c(Y, Z)$. If $y \in Y$, then $f(?, y)$ is continuous from X to Z, so it sends the bounded set A to a bounded subset of Z. That is, $\{f(a, y) : a \in A\}$ is bounded in Z whenever $y \in Y$. By Proposition 4.14(a), this just says that $\mathscr{F}$ is bounded for pointwise convergence, so $\mathscr{F}$ is equicontinuous by Theorem 4.16(d).

Suppose a net $\langle (x_\alpha, y_\alpha) \rangle$ converges to (x, y) in $A \times Y$, and suppose W is a convex, balanced neighborhood of 0 in Z. Then there is a neighborhood V of 0 in Y such that $f(a, V) \subset \frac{1}{2}W$ for all $a \in A$ (equicontinuity). Also, there is a neighborhood U of 0 in X such that $f(U, y) \subset \frac{1}{2}W$ since $f(?, y)$ is continuous. Finally, there is an α such that $\beta \succ \alpha \Rightarrow (x_\beta, y_\beta) \in (x, y) + U \times V$. From this, for $\beta \succ \alpha$:

$$f(x_\beta, y_\beta) - f(x, y) = f(x_\beta, y_\beta) - f(x_\beta, y) + f(x_\beta, y) - f(x, y)$$
$$= f(x_\beta, y_\beta - y) + f(x_\beta - x, y)$$
$$\in \underbrace{\frac{1}{2}W}_{\substack{\text{since } x_\beta \in A \\ \text{and } y_\beta - y \in V}} + \underbrace{\frac{1}{2}W}_{\substack{\text{since} \\ x_\beta - x \in U}} = W.$$

That is, $\langle f(x_\alpha, y_\alpha) \rangle \to f(x, y)$. Hence f is continuous [Proposition 1.3(c)]. □

Corollary 4.18. *Suppose X, Y, and Z are Hausdorff locally convex spaces, and $f : X \times Y \to Z$ is a separately continuous bilinear map. Suppose X is first countable and Y is barreled. Then f is jointly continuous provided either X is normed or Y is a Fréchet space.*

Proof. Case 1: X is normed. Let A_n be the open ball in X of radius n. Then f is continuous on $A_n \times Y$. Thus if W is open in Z, then $f^{-1}(W) \cap (W_n \times Y)$ is relatively open in $A_n \times Y$ since A_n is bounded in X. But $A_n \times Y$ is open in $X \times Y$, so $f^{-1}(W) \cap (A_n \times Y)$ is open in $X \times Y$. Hence

$$f^{-1}(W) = \bigcup_{n=1}^{\infty} [f^{-1}(W) \cap (A_n \times Y)]$$

is open in $X \times Y$.

Case 2: Y is a Fréchet space. Then $X \times Y$ is first countable, so sequences suffice to check continuity. If $(x_n, y_n) \to (x, y)$ in $X \times Y$, then $x_n \to x$ in X and $y_n \to y$ in Y, so $A = \{x\} \cup \{x_n : n \in \mathbb{N}\}$ is compact, hence bounded, in X. Since f is continuous on $A \times Y$, $f(x_n, y_n) \to f(x, y)$. □

There is one situation where a separately continuous bilinear map arises naturally, and it illustrates the limits to generalizing Corollary 4.18: the evaluation map. Suppose Y is a Hausdorff locally convex space, and $X = Y^*$, its strong dual. Let $Z = \mathbb{F}$ be the base field, and define $e : Y^* \times Y \to \mathbb{F}$ by $e(f, y) = f(y)$. $e(f, ?) = f$, which is continuous; while $e(?, y)$ has values of modulus $\leq \varepsilon$ on $\{\varepsilon^{-1} y\}^\circ$, and so is also continuous. That is, the evaluation map is *always* separately continuous. However, it is not jointly continuous unless Y can be normed; see Exercise 5. If Y is an LB-space, then Y^* is a Fréchet space (Proposition 3.46), and so is first countable, while Y itself is barreled (Corollary 4.6). This shows that something beyond "X is first countable and Y is barreled" is necessary for Corollary 4.18. Incidentally, it also shows that the strong dual of a Fréchet space Y

cannot be another Fréchet space unless Y can actually be normed. It cannot even be first countable ... but (thanks to results discussed in the next section) the strong dual of a Fréchet space is always complete.

4.3 Completeness

When working with Banach spaces, or even normed spaces, $\mathscr{L}_c(X, Y)$ is complete whenever Y is complete. However, other things also happen automatically. For example, bounded linear transformations are always continuous when working with normed spaces, a result that fails when X is locally convex and not bornological. *These two facts are interrelated.* A digression is in order, concerning the space $\mathscr{L}_b(X, Y)$ of bounded linear transformations from X to Y.

Technically, the first point concerns the vector space of all linear transformations, but it is easily deduced from first principles (plus one result from Appendix A).

Proposition 4.19. *Suppose X and Y are Hausdorff topological vector spaces, and $\langle T_\alpha \rangle$ is a net consisting of (not necessarily continuous) linear transformations from X to Y. Suppose that for all $x \in X$, $\lim T_\alpha(x)$ exists: Call that limit $T(x)$. Then T is a linear transformation.*

Proof. If $x \in X$ and c is in the base field, then

$$T(cx) = \lim T_\alpha(cx) = \lim c T_\alpha(x) = c \lim T_\alpha(x) = c T(x)$$

since multiplication by c is continuous. Similarly, if $x, y \in X$, then $T_\alpha(x) \to T(x)$ and $T_\alpha(y) \to T(y)$ in Y, so $(T_\alpha(x), T_\alpha(y)) \to (T(x), T(y))$ in $Y \times Y$ (Proposition A.2). Hence $T_\alpha(x) + T_\alpha(y) \to T(x) + T(y)$ since addition is jointly continuous. Thus,

$$T(x + y) = \lim T_\alpha(x + y) = \lim T_\alpha(x) + T_\alpha(y) = T(x) + T(y). \qquad \square$$

Now back to $\mathscr{L}_b(X, Y)$. If A is bounded in X and U is a neighborhood of 0 in Y, set

$$N_b(A, U) = \{T \in \mathscr{L}_b(X, Y) : T(A) \subset U\}.$$

The subscript "b" is included to emphasize the fact that we are dealing with $\mathscr{L}_b(X, Y)$ and not $\mathscr{L}_c(X, Y)$. Note that if $T(A)$ is bounded, say $T(A) \subset cU$, then $T \in N_b(A, cU) = cN_b(A, U)$ [Lemma 4.13, which is a computation, and applies to $N_b(A, U)$ as well as $N(A, U)$]. That is, the sets $N_b(A, U)$ are absorbent, and (as in Sect. 3.6) form a neighborhood base of 0 for the topology of bounded convergence on $\mathscr{L}_b(X, Y)$.

Theorem 4.20. *Suppose X and Y are Hausdorff locally convex spaces, and suppose Y is complete. Then in the topology of bounded convergence, $\mathscr{L}_b(X, Y)$, the space of bounded linear transformations from X to Y, is complete.*

Proof. Suppose $\langle T_\alpha \rangle$ is a Cauchy net in $\mathscr{L}_b(X, Y)$. If $x \in X$, then

$$T_\alpha - T_\beta \in N_b(\{x\}, U) \Rightarrow T_\alpha(x) - T_\beta(x) \in U$$

which happens when α and β are both "large," so for all $x \in X$ $\langle T_\alpha(x) \rangle$ is a Cauchy net in Y. Hence $T(x) = \lim T_\alpha(x)$ exists for all $x \in X$ since Y is complete. This T is linear by Proposition 4.19; it remains to show that T is bounded and $T_\alpha \to T$ in $\mathscr{L}_b(X, Y)$. These are done simultaneously.

Suppose A is bounded in X and U is a neighborhood of 0 in Y. Choose a barrel neighborhood V of 0 in Y such that $V \subset U$. there exists α s.t. $\beta, \gamma \succ \alpha \Rightarrow T_\beta - T_\gamma \in N_b(A, V)$. Thus, if $x \in A$, then $\beta, \gamma \succ \alpha \Rightarrow T_\beta(x) - T_\gamma(x) \in V$. But $T_\gamma(x) \to T(x)$; taking the limit in γ (Proposition 1.4: $\{\gamma : \gamma \succ \alpha\}$ is cofinal) gives $T_\beta(x) - T(x) \in V$ since V is closed, when $\beta \succ \alpha$ and $x \in A$.

This has two consequences. First, setting $\beta = \alpha$ and eventually letting A and U float, $T_\alpha(x) - T(x) \in V$, so $T(x) - T_\alpha(x) \in V$ (V is balanced), and there exists $c > 0$ such that $T_\alpha(A) \subset cV$. Hence for all $x \in A$:

$$T(x) = T(x) - T_\alpha(x) + T_\alpha(x) \in V + cV = (c + 1)V \subset (c + 1)U$$

[Proposition 3.3(b)]. That is, $T(A) \subset (c + 1)U$. With U varying: $T(A)$ is bounded. With A varying: T is bounded.

Now that we know that T is bounded, $T_\beta(x) - T(x) \in V$ when $\beta \succ \alpha$ and $x \in A$ means that $(T_\beta - T)(A) \subset V$ when $\beta \succ \alpha$, that is $T_\beta - T \in N_b(A, V) \subset N_b(A, U)$ when $\beta \succ \alpha$. *This* means that $T_\alpha \to T$ in the topology of bounded convergence. $\square$

Corollary 4.21. *Suppose X and Y are Hausdorff locally convex spaces, and suppose X is bornological and Y is complete. Then $\mathscr{L}_c(X, Y)$ is complete in the topology of bounded convergence.*

Proof. Since X is bornological, $\mathscr{L}_c(X, Y) = \mathscr{L}_b(X, Y)$ (Theorem 4.12) and $\mathscr{L}_b(X, Y)$ is complete (Theorem 4.20). $\square$

Corollary 4.22. *The strong dual of a Hausdorff, bornological, locally convex space is complete.*

Proof. The base field is complete. $\square$

By the way, there are nonbornological spaces with duals that are complete, but more discussion of duality is needed for this.

If "completeness" alone were all we ever needed, we would now be done with this section. However, there has been no discussion here of sequential completeness. It turns out that sequential completeness is not the "right" concept here, although there are general results in the subject; see Exercise 10, for example. The reason is that an intermediate concept really needs to be introduced.

Definition 4.23. Suppose X is a Hausdorff locally convex space. Then X is **quasi-complete** if every bounded Cauchy net is convergent.

Observe that complete implies quasi-complete trivially. Also, quasi-complete implies sequentially complete, since every Cauchy sequence is bounded: If $\langle x_n \rangle$ is a Cauchy sequence, and U is a convex, balanced neighborhood of 0, then there exists N s.t. $m, n \geq N \Rightarrow x_m - x_n \in U$. Also, there exists $c > 0$ such that $\{x_1, \ldots, x_N\} \subset cU \subset (c+1)U$. But then $x_n = x_n - x_N + x_N \in U + cU = (c+1)U$ when $n > N$, too, so all $x_n \in (c+1)U$. (*Note:* A generality below, Theorem 4.28, will also cover this.)

Theorem 4.24. *Suppose X and Y are Hausdorff locally convex spaces, and suppose Y is quasi-complete.*

(a) *If X is infrabarreled, then $\mathscr{L}_c(X, Y)$ is quasi-complete in the topology of bounded convergence.*

(b) *If X is barreled, then $\mathscr{L}_c(X, Y)$ is quasi-complete in the topology of pointwise convergence.*

Proof. Suppose $\langle T_\alpha \rangle$ is a Cauchy net that is bounded in the topology of pointwise convergence. Then for all $x \in X$, $\{T_\alpha(x)\}$ is bounded by Proposition 4.14(a). Also: $T_\alpha - T_\beta \in N(\{x\}, U) \Leftrightarrow T_\alpha(x) - T_\beta(x) \in U$, and the former can be enforced by requiring α and β to be "large," so (letting U vary) $\langle T_\alpha(x) \rangle$ is a Cauchy net in Y. Hence $T(x) = \lim T_\alpha(x)$ exists since Y is quasi-complete, and this T is a linear transformation (Proposition 4.19). The point here is that in both (a) and (b), we have a limiting transformation. (Note that a bounded Cauchy net in the topology of bounded convergence is a bounded Cauchy net in the topology of pointwise convergence.)

Now suppose A is bounded [part (a)] or finite [part (b)]. If U is a neighborhood of 0 in Y, choose a barrel neighborhood V of 0 in Y such that $V \subset U$. Then as in the proof of Theorem 4.20, two things happen, but the reasoning now is different.

First of all, by Theorem 4.16, $\{T_\alpha\}$ is equicontinuous [Theorem 4.16(b) for part (a), Theorem 4.16(d) for part (b)]. Hence there exists a neighborhood W of 0 in X such that for all $x \in W$ and all α, $T_\alpha(x) \in V$. Hence $T(x) = \lim T_\alpha(x) \in V$ when $x \in W$ since V is closed. That is, $W \subset T^{-1}(V)$. Letting V float, T is now continuous by Proposition 1.26.

Finally, $T_\alpha \to T$ in the relevant topology: there exists α s.t. $\beta, \gamma \succ \alpha \Rightarrow T_\beta - T_\gamma \in N(A, V)$. That is, if $x \in A$, then $T_\beta(x) - T_\gamma(x) \in V$ when $\beta, \gamma \succ \alpha$. Again, letting $\gamma \to \infty$, $T_\beta(x) - T(x) \in V$ since V is closed (this part follows the pattern in Theorem 4.20), when $\beta \succ \alpha$ and $x \in A$, so that $T_\beta - T \in N(A, V) \subset N(A, U)$ when $\beta \succ \alpha$. This just means that $T_\alpha \to T$ in the relevant topology, by letting A and U vary. $\qquad \square$

Corollary 4.25. *Suppose X is a Hausdorff locally convex space.*

(a) *If X is infrabarreled, then X^* is quasi-complete in the strong topology.*

(b) *If X is barreled, then X^* is quasi-complete in the weak-* topology.*

Part (b) in both results is a bit of a surprise; quasi-completeness does not require that the topology be excessively fine. However, one's intuitive association of completeness with fineness is not totally wrong; see Exercise 9. Completeness definitely does not always descend to coarser topologies: A nonreflexive Banach space is not even quasi-complete in its weak topology (Chap. 5, Exercise 13). By the way, the weak-* topology on X^* cannot actually be complete unless $X^* = X'$; see Exercise 4 from Chap. 3. However, there are spaces for which $X^* = X'$ (Example I from Sect. 3.8) and for which X^* is complete in the weak-* topology (see Exercise 11), so even completeness does not require a space to have a spectacularly fine topology.

Quasi-completeness is a useful concept. It implies sequential completeness, but it also has application beyond that. Probably the most important concerns the relation between compactness and precompactness.

Definition 4.26. Suppose X is a Hausdorff locally convex space, and $A \subset X$. A is **precompact** if for each neighborhood U of 0, there exists $x_1, \ldots, x_n \in A$ such that $A \subset \cup x_j + U$.

This formulation will help later. Many texts use the phrase "totally bounded" instead of precompact. Note that the definition makes sense in any topological vector space (or even in a topological group, for "left precompact"). However, it is for Hausdorff locally convex spaces that the generalities in Theorem 4.28 will be easily proven. We need a lemma first.

Lemma 4.27. *Suppose X is a Hausdorff locally convex space, and $A \subset X$. Then the following are equivalent:*

(i) A is precompact.
(ii) For each neighborhood U of 0, there exists a finite set $F_U \subset X$ such that $A \subset F_U + U$.
(iii) For each neighborhood U of 0, there exists a compact set $K_U \subset X$ such that $A \subset K_U + U$.

Proof. (i) $\Rightarrow$ (ii), since the definition provides $F_U = \{x_1, \ldots, x_n\} \subset A$. Then (ii) $\Rightarrow$ (iii), since finite sets are compact. Suppose (iii). Given a neighborhood U of 0, choose an open convex, balanced neighborhood V of 0 for which $V \subset \frac{1}{4}U$, that is $4V \subset U$. Then $A \subset K_V + V$. Now K_V is covered by all $y + V$, $y \in K_V$, so there exists $y_1, \ldots, y_N$ in K_V for which $K_V \subset \cup y_j + V$. Reorder so that $A \cap (y_j + 2V) \neq \emptyset$ for $j = 1, \ldots, n$, and $A \cap (y_j + 2V) = \emptyset$ for $j = n + 1, \ldots, N$. Choose $x_j \in A \cap (y_j + 2V)$ for $j = 1, \ldots, n$. The claim is that $A \subset \cup x_j + 4V \subset \cup x_j + U$.

Suppose $a \in A$. Then $a \in A \subset K_V + V$, so there exists $x \in K_V$ and $v \in V$ such that $a = x + v$, that is $a - v = x$. But $x \in y_j + V$ for some j. Hence $a = x + v \in y_j + V + V = y_j + 2V$ [Proposition 3.3(b)], so $A \cap (y_j + 2V)$ is nonempty, and $j \leq n$. Write $a = y_j + w$, $w \in 2V$. Also, $x_j \in y_j + 2V$, so $y_j = x_j + w'$, $w' \in 2V$, since $2V$ is balanced. But now $a = x_j + w' + w \in x_j + 2V + 2V = x_j + 4V$. $\square$

Theorem 4.28. *Suppose X is a Hausdorff locally convex space.*

(a) *Compact sets are precompact.*
(b) *Precompact sets are bounded.*
(c) *Cauchy sequences are precompact.*
(d) *The closure of a precompact set is precompact.*
(e) *Any subset of a precompact set is precompact.*
(f) *The union of finitely many precompact sets is precompact.*
(g) *The convex hull of a precompact set is precompact.*
(h) *A complete precompact set is compact.*

Proof. (a) Is trivial.

(b) If A is precompact and U is a convex, balanced neighborhood of 0, then $A \subset K_U + U$ for some compact K_U. K_U is bounded, so $K_U \subset cU$ for some $c > 0$, so $A \subset K_U + U \subset cU + U = (c + 1)U$.

(c) If $\langle x_n \rangle$ is a Cauchy sequence and U is any neighborhood of 0, there exists N s.t. $x_n - x_m \in U$ when $n, m \geq N$. In particular, $x_n - x_N \in U$, that is $x_n \in x_N + U$, when $n \geq N$. Hence $x_1 + U, \ldots, x_N + U$ covers $\{x_n\}$.

(d) If A is precompact, and U is a neighborhood of 0, choose a convex, balanced neighborhood V of 0 with $V \subset \frac{1}{2}U$. Then $A \subset K_V + V$, so by Proposition 1.9, $A^- \subset A + V \subset K_V + V + V = K_V + 2V \subset K_V + U$.

(e) If $A \subset K_U + U$ and $B \subset A$, then $B \subset K_U + U$, too.

(f) If each $A_j \subset K_{U,j} + U$, then $\bigcup_{j=1}^n A_j$ is contained in $\bigcup_{j=1}^n K_{U,j} + U$.

(g) Suppose A is precompact and U is a neighborhood of 0. Choose an open, convex, balanced neighborhood V of 0 for which $V \subset U$. Choose $x_1, \ldots, x_n \in A$ for which $A \subset \bigcup x_j + V$. Set $K_U =$ convex hull of $\{x_1, \ldots, x_n\}$. Then

$$K_U = \left\{ \sum_{j=1}^n t_j x_j : \text{all } t_j \geq 0 \text{ and } \Sigma t_j = 1 \right\}$$

is the continuous image of a compact simplex in $\mathbb{R}^n$, so K_U is compact. Now K_U is also convex, as is V, so $A \subset K_U + V$, a convex set. Hence con(A), the convex hull of A, is contained in $K_U + V \subset K_U + U$.

(h) Suppose not; suppose A is precompact and complete but is not compact. Following Proposition A.13 in Appendix A, let $\mathscr{P}$ be an open cover of A which is maximal (under set inclusion) with respect to the property of not having a finite subcover of A. Then $\emptyset \in \mathscr{P}$, and if U and V are open with $U \cap V \in \mathscr{P}$, then $U \in \mathscr{P}$ or $V \in \mathscr{P}$ by Proposition A.13. Now let D be the directed set of open, convex, balanced neighborhoods of 0 in X, where $U \succ V$ when $U \subset V$. If $U \in D$, then since A is precompact, $A \subset F + U$ for some finite set F in A. (This is where the requirement in the definition of precompact that all $x_j \in A$ is helpful.) Now $\{x + U : x \in F\}$ is a finite cover of A, so it is not a subcover of $\mathscr{P}$, so there exists $x_U \in F$ for which $x_U + U \notin \mathscr{P}$. The claim is that $\langle x_U \rangle$ is a Cauchy net, after which a contradiction will be derived from the fact that its limit must be covered by $\mathscr{P}$.

$\langle x_U \rangle$ is a Cauchy net: Suppose U is an open, convex, balanced neighborhood of 0. If $V, W \in D$ and $V, W \subset \frac{1}{2}U$, then $x_V + V$ and $x_W + W$ are not in $\mathscr{P}$, so $(x_V + V) \cap (x_W + W) \notin \mathscr{P}$. Since $\emptyset \in \mathscr{P}$, $(x_V + V) \cap (x_W + W) \neq \emptyset$, so there exists $x \in (x_V + V) \cap (x_W + W)$. Hence $x - x_V \in V$ and $x - x_W \in W$, so that $x_V - x \in V$ (V is balanced), and $x_V - x_W = x_V - x + x - x_W \in V + W \subset \frac{1}{2}U + \frac{1}{2}U = U$.

A is complete; set $x = \lim x_U$. Choose $W \in \mathscr{P}$ for which $x \in W$. Choose $U_1 \in D$ for which $x + U_1 \subset W$. Choose $U_2 \in D$ for which $V \subset U_2 \Rightarrow x_V \in x + \frac{1}{2}U_1$. Set $U = \left(\frac{1}{2}U_1\right) \cap U_2$; then $U \in D$, and $U \subset U_2$, so $x_U \in x + \frac{1}{2}U_1$. But $U \subset \frac{1}{2}U_1$ as well, so $x_U + U \subset x + \frac{1}{2}U_1 + \frac{1}{2}U_1 = x + U_1 \subset W$. Hence $x_U + U \in \mathscr{P}$ by Proposition A.13, since $W \in \mathscr{P}$, contradicting the choice of x_U.   □

Corollary 4.29. *Suppose X is a Hausdorff, quasi-complete, locally convex space. Then the closed convex hull of a compact set is compact.*

Proof. If A is compact, then A is precompact [part (a)], hence $\text{con}(A)$ is precompact [part (g)], hence $\text{con}(A)^-$ is precompact [part (d)] and complete [part (b)], hence is compact [part (h)].   □

Corollary 4.30. *Suppose X is a Hausdorff, quasi-complete, locally convex space. If A is compact in X, then $(A^\circ)_\circ$ is compact.*

Proof. Let K denote the closed convex hull of A; then K is compact by Corollary 4.29. Hence $(K^\circ)_\circ$ is compact by Proposition 3.21. But $A \subset K \Rightarrow A^\circ \supset K^\circ \Rightarrow (A^\circ)_\circ \subset (K^\circ)_\circ$ by Proposition 3.19, so $(A^\circ)_\circ$ is compact since it is also closed.   □

Remark. $(A^\circ)_\circ = (K^\circ)_\circ$ in the above proof; see Exercise 12.

One final note. For Corollaries 4.29 and 4.30, what is really needed as a property of X is that all precompact Cauchy nets are convergent. However, this does not seem to be very helpful, since precompact sets are hard to identify functional analytically, whereas Corollary 3.31 identifies bounded sets via the dual.

4.4 The Open Mapping Theorem

The proof of the open mapping theorem for Banach spaces has two parts: a category argument in the range space, followed by a sequential argument in the domain. The proof here will follow a similar pattern, but it is useful to make note of what is known at the halfway point. This involves a new concept which is actually topological in nature.

Definition 4.31. Suppose X and Y are topological spaces, and $f : X \to Y$ is a function. Then f is **nearly open** if for all open $U \subset X$:

$$f(U) \subset \text{int}(f(U)^-).$$

Here, we are primarily concerned with continuous, nearly open maps. The following facts appear among the exercises at the end of this chapter:

1. Suppose Y is a topological space, and X is a dense subspace. Then the inclusion map $X \hookrightarrow Y$ is nearly open.
2. Suppose X is a locally compact Hausdorff space and Y is a Hausdorff space. Then any continuous, nearly open map $f : X \to Y$ is an open map.
   This seems to indicate that some form of completeness and uniformity is needed to force a continuous, nearly open map to be open. Completeness alone, however, is not enough.
3. There exists a complete metric space X, a compact metric space Y, and a continuous (Lipschitz, actually), nearly open map $f : X \to Y$ which is not open.

Here, as sets,

$$X = \bigcup_{n=1}^{\infty} [0, 1] \cdot e^{i\pi/n} \text{ in } \mathbb{C} \approx \mathbb{R}^2, \text{ and}$$

$$Y = X^- \qquad\qquad f(z) = z$$

The metric on Y is the usual Euclidean metric on $\mathbb{C}$, while the metric on X is the "Washington metric" discussed in the second "Nonexample" at the beginning of Sect. 2.1: $d(z, w) = |z - w|$ if z and w are colinear, but $d(z, w) = |z| + |w|$ if not.

The next result will be useful for establishing 1–3 above, and illuminate the situation for linear maps on locally convex spaces. (Linearity will eventually provide the needed uniformity.)

Proposition 4.32. *Suppose X and Y are topological spaces, and $f : X \to Y$ is a function. Then the following are equivalent:*

(i) f is nearly open.
(ii) There exists a global base $\mathscr{B}$ for the topology on X (consisting of open sets) for which $f(B) \subset \text{int}(f(B)^-)$ for all $B \in \mathscr{B}$.
(iii) For all $x \in X$, there exists a local base $\mathscr{B}_x$ for the topology on X (not necessarily consisting of open sets) for which $f(x) \in \text{int}(f(B)^-)$ for all $B \in \mathscr{B}_x$.

Proof. (i) $\Rightarrow$ (ii) Since global bases exist (the whole topology is one such!), and global bases consist of open sets B for which $f(B) \subset \text{int}(f(B)^-)$ when f is nearly open.

(ii) $\Rightarrow$ (iii) Since if $\mathscr{B}$ is a global base as in (ii), then $\mathscr{B}_x = \{B \in \mathscr{B} : x \in B\}$ is a local base for which $f(x) \in f(B) \subset \text{int}(f(B)^-)$.

(iii) $\Rightarrow$ (i) Suppose U is open in X, and $x \in U$. Let $\mathscr{B}_x$ be as in (iii) Choose $B \in \mathscr{B}_x$ for which $B \subset U$. Then $f(B) \subset f(U)$, $f(B)^- \subset f(U)^-$, and $f(x) \in \text{int}(f(B)^-) \subset \text{int}(f(U)^-)$. Since $f(x)$ is actually an arbitrary point in $f(U)$, $f(U) \subset \text{int}(f(U)^-)$.

$\square$

Corollary 4.33. *Suppose X and Y are topological vector spaces, and $T : X \to Y$ is a linear map. Let $\mathscr{B}_0$ denote a base for the topology of X at 0. Then T is nearly open if, and only if, $T(B)^-$ is a neighborhood of 0 in Y for all $B \in \mathscr{B}_0$.*

Proof. If part: for all $x \in X$, $x + \mathscr{B}_0$ is a base at x, and for all $B \in \mathscr{B}_0 : 0 \in \text{int}(T(B)^-) \Rightarrow T(x) \in T(x) + \text{int}(T(B)^-) = \text{int}(T(x) + T(B)^-) = \text{int}(T(x + B)^-)$ since translation is a homeomorphism.

Only if part: $0 \in T(\text{int}(B)) \subset \text{int}(T(\text{int}(B))^-) \subset \text{int}(T(B)^-)$ for all $B \in \mathscr{B}_0$ when T is nearly open. □

Corollary 4.34. *Suppose X and Y are Hausdorff locally convex spaces, and suppose Y is barreled. Then any linear map T from X onto Y is nearly open.*

Proof. Use $\mathscr{B}_0 =$ all convex, balanced neighborhoods of 0 in X. If $B \in \mathscr{B}_0$, and $x \in X$, then $x \in cB \Rightarrow T(x) \in T(cB) = cT(B)$, so $T(B)$ is convex, balanced, and absorbent (since T is onto). Hence $T(B)^-$ is closed, convex (Proposition 2.13), balanced (Proposition 2.5), and absorbent, that is $T(B)^-$ is a barrel in Y. Since Y is assumed to be barreled, $T(B)^-$ is a neighborhood of 0. Hence T is nearly open by Corollary 4.33. □

(There are certain points where using the barreled condition seems almost like cheating. This is one of them.)

We can now prove the open mapping theorem in our Hausdorff, locally convex space context.

Theorem 4.35 (Open Mapping Theorem). *Suppose X and Y are Hausdorff locally convex spaces, and $T : X \to Y$ is a continuous linear map.*

(a) If T is onto and Y is barreled, then T is nearly open.
(b) If T is nearly open and X is a Fréchet space, then T is open (and onto).

Proof. Part (a) is immediate from Corollary 4.34. For part (b), assume T is nearly open and X is a Fréchet space. Let U be a neighborhood of 0 in X. Choose a base $\mathscr{B}_0 = \{B_1, B_2, \ldots\}$ in accordance with Theorem 1.13 for which $B_j = -B_j$ and $B_{j+1} + B_{j+1} \subset B_j$, all B_j being closed, and $B_1 \subset U$.

Suppose $y \in T(B_2)^-$. Then since $T(B_3)^-$ is a neighborhood of 0,

$$y \in T(B_2)^- \subset T(B_2) + T(B_3)^-,$$

so $y = T(x_2) + y_3$, $x_2 \in B_2$ and $y_3 \in T(B_3)^-$. Since $T(B_4)^-$ is a neighborhood of 0,

$$y_3 \in T(B_3)^- \subset T(B_3) + T(B_4)^-, \text{ so}$$

$$y_3 = T(x_3) + y_4; x_3 \in B_3 \text{ and } y_4 \in T(B_4)^-.$$

Recursively:

$$y_n \in T(B_n)^- \subset T(B_n) + T(B_{n+1})^-, \text{ so}$$

$$y_n = T(x_n) + y_{n+1}, x_n \in B_n \text{ and } y_{n+1} \in T(B_{n+1})^-.$$

We now have that $x = \Sigma x_j$ converges, thanks to Theorem 1.35. Now, as a result of the observation in the paragraph preceding Theorem 1.35, $B_2 + B_3 + \cdots + B_n \subset B_1$, that is all partial sums of Σx_j lie in B_1, so $x \in B_1$ since B_1 is closed. There is more:

$$y = T(x_2) + y_3 = T(x_2) + T(x_3) + y_4$$
$$= T(x_2) + T(x_3) + T(x_4) + y_5$$
$$= \cdots = \left(\sum_{j=2}^{n} T(x_j) \right) + y_{n+1} = T \left(\sum_{j=2}^{n} x_j \right) + y_{n+1}$$

So:

$$y_{n+1} = y - T \left(\sum_{j=2}^{n} x_j \right) \to y - T(x)$$

since T is continuous. Note that $y_n \in T(B_n)^-$, and

$$k \geq n \Rightarrow B_k \subset B_n \Rightarrow y_k \in T(B_k)^- \subset T(B_n)^-,$$

so $y - T(x) \in T(B_n)^-$ for all n. Now if V is any closed neighborhood of 0 in Y, then $B_n \subset T^{-1}(V)$ for some n since T is continuous, so $T(B_n) \subset V$ and $y - T(x) \in T(B_n)^- \subset V$ since V is closed. Hence $y - T(x) = 0$ since Y is Hausdorff. We now have that $y = T(x) \in T(B_1) \subset T(U)$. Since y was arbitrary in $T(B_2)^-$: $T(B_2)^- \subset T(U)$, so $T(U)$ is a neighborhood of 0 in Y. Hence T is open by Proposition 1.26(b). Finally, T is now onto, since $T(X)$ must be an open subspace of Y, that is $T(X) = Y$. □

Remark. The above proof that "nearly open $\Rightarrow$ open" actually works for topological groups. The hypotheses required are that X and Y be Hausdorff topological groups, with X being first countable and complete; and $T : X \to Y$ be a continuous, nearly open homomorphism. The conclusion is that T is an open map, and that $T(X)$ is then an open subgroup of Y. The proof is almost identical: All you have to do is replace each plus sign with a multiplication symbol (e.g., "$y = T(x_2) + y_3$" becomes "$y = T(x_2) \cdot y_3$, and "$x = \Sigma x_j$" becomes "$x = \prod x_j$"); after reversing the order of $y - T(x)$ [and $y - T(\Sigma x_j)$], which, for example, becomes $-T(x) + y$ and then $T(x)^{-1} \cdot y$. That's it. Something similar happens with the closed graph theorem.

By the way, a typical application of the "nearly open $\Rightarrow$ onto" part will appear in Chap. 5.

Corollary 4.36. *Suppose X is a Fréchet space, and Y is a barreled, Hausdorff, locally convex space. Suppose $T : X \to Y$ is a continuous linear map from X onto Y. Then T induces an isomorphism of the Fréchet space $X/\ker(T)$ with Y.*

Proof. The induced map is continuous and open by Theorem 1.23(c) and (e), and is an algebraic isomorphism for the usual algebraic reasons. $X/\ker(T)$ is Hausdorff since $\ker T = T^{-1}(\{0\})$ is closed [Theorem 1.23(g)], while $X/\ker(T)$ is first countable by Theorem 1.23(f) and complete by Corollary 1.36. Hence $X/\ker(T)$ is a Fréchet space (Corollary 3.36). $\qquad\qquad\qquad\qquad\qquad\qquad\qquad\qquad\qquad\qquad$ □

Now for the closed graph theorem.

4.5   The Closed Graph Theorem

For Banach spaces, it is traditional to prove the open mapping theorem, and then derive the closed graph theorem as a corollary. There is a reason for that, even though the open mapping theorem can be just as easily derived from the closed graph theorem. A direct proof of the open mapping theorem involves two steps; these are parts (a) and (b) of Theorem 4.35 in the last section. [In our context, part (a) was trivial only because Theorem 4.5 was available.] A direct proof of the closed graph theorem, however, involves three steps. Here, we have to go through that because the two results are largely independent. True, one can derive Corollary 4.36 from the closed graph theorem as it appears here, but Theorem 4.35b) does not directly follow from this. Also, the closed graph theorem cannot be derived directly from the open mapping theorem due to the asymmetry in the conditions on the spaces: The graph does not inherit any nice properties.

There *is* a version of the closed graph theorem that can be used to directly prove the open mapping theorem; it is rather messy, and is given in Appendix B. (It applies to topological groups.)

Theorem 4.37 (Closed Graph Theorem). *Suppose X and Y are Hausdorff locally convex spaces, and suppose X is barreled and Y is a Fréchet space. Suppose $T : X \to Y$ is a linear transformation with a graph, $\Gamma(T)$, that is closed in $X \times Y$. Then T is continuous.*

Proof. Let U be a convex, balanced neighborhood of 0 in Y. In view of Proposition 1.26(a), it suffices to prove that for any such U, $T^{-1}(U)$ is a neighborhood of 0. As in the proof of the open mapping theorem, choose a base at 0 for Y : $\mathscr{B}_0 = \{B_1, B_2, \ldots\}$, with $B_j = -B_j$, $B_{j+1} + B_{j+1} \subset B_j$, all B_j closed, and $B_1 \subset U$. Given any B_j, there exists a convex, balanced neighborhood W of 0 such that $W \subset B_j$. Now given any x, W absorbs $T(x)$, so $T^{-1}(W)$ absorbs x. That is, $T^{-1}(W)$ is convex, balanced, and absorbent, so $T^{-1}(W)^-$ is a barrel, and so is a neighborhood of 0 in X. Since $W \subset B_j$: $T^{-1}(W) \subset T^{-1}(B_j)$, so $T^{-1}(W)^- \subset T^{-1}(B_j)^-$. That is, every $T^{-1}(B_j)^-$ is a neighborhood of 0 in X. This is the first step in the proof.

For the second step, suppose $x \in T^{-1}(B_2)^-$. We will eventually show that $T(x) \in U$, so that $x \in T^{-1}(U)$, giving $T^{-1}(B_2)^- \subset T^{-1}(U)$ by letting x vary. This will complete the proof, but there are two distinct parts to this. The current one produces a candidate $y \in B_1 \subset U$ for which (eventually) $T(x)$ will equal y, and the last step will establish that $T(x) = y$. This y is the sum of a series in the same manner as occurred in the proof of the open mapping theorem. Now $x \in T^{-1}(B_2)^-$, so again

$$x \in T^{-1}(B_2)^- \subset T^{-1}(B_2) + T^{-1}(B_3)^-$$

by Proposition 1.9. Write $x = x_2 + x_3'$, with $T(x_2) \in B_2$ and $x_3' \in T^{-1}(B_3)^-$. In general, given

$$x_n' \in T^{-1}(B_n)^- \subset T^{-1}(B_n) + T^{-1}(B_{n+1})^-$$

recursively set $x_n' = x_n + x_{n+1}'$, $x_n \in T^{-1}(B_n)$ and $x_{n+1}' \in T^{-1}(B_{n+1})^-$. Hence $T(x_n) \in B_n$. In general,

$$x = x_2 + x_3' = x_2 + x_3 + x_4' = \cdots = \sum_{j=2}^{n} x_j + x_{n+1}'.$$

so that $x - \sum_{j=2}^{n} x_j = x_{n+1}' \in T^{-1}(B_{n+1})^-$.

Since $T(x_j) \in B_j$, $y = \sum T(x_j)$ converges in Y by Theorem 1.35.

The final step is to show that $T(x) = y$. This is where we use the fact that $\Gamma(T)$ is closed. To show that $T(x) = y$, it suffices to show that if V is a convex, balanced neighborhood of 0 in X and n is any index, then

$$[(x, y) + V \times B_{n-1}] \cap \Gamma(T) \neq \emptyset,$$

since (x, y) will then be adherent to $\Gamma(T)$, hence $(x, y) \in \Gamma(T)$ since $\Gamma(T)$ is closed. For this purpose, note that

$$x - \sum_{j=2}^{n} x_j = x_{n+1}' \in T^{-1}(B_{n+1})^- \subset T^{-1}(B_{n+1}) + V$$

so that $x - \sum_{j=2}^{n} x_j = x_{n+1}'' + v$, $T(x_{n+1}'') \in B_{n+1}$, $v \in V$.

and $x_{n+1}'' + \sum_{j=2}^{n} x_j = x - v \in x + V$.

Now

$$y = \sum_{j=2}^{\infty} T(x_j) = \sum_{j=2}^{n} T(x_j) + \sum_{j=n+1}^{\infty} T(x_j)$$

$$\text{set } \tilde{y}_n = \sum_{j=n+1}^{\infty} T(x_j), \text{ so that } y = \tilde{y}_n + T\left(\sum_{j=2}^{n} x_j\right).$$

Now $B_{n+1} + B_{n+2} + \cdots + B_N \subset B_n$ as in the discussion preceding Theorem 1.35, so the partial sums for $\tilde{y}_n$ are in B_n. Hence $\tilde{y}_n \in B_n$ since B_n is closed. Now $T(x_{n+1}'') \in B_{n+1}$, so $T(x_{n+1}'') - \tilde{y}_n \in B_{n+1} + B_n \subset B_{n-1}$ $(B_n = -B_n)$. Hence

$$T\left(x_{n+1}'' + \sum_{j=2}^{n} x_j\right) = y + T(x_{n+1}'') - \tilde{y}_n$$

$$\in y + B_{n-1}.$$

Thus (finally!):

$$(x, y) + (-v, T(x_{n+1}'') - \tilde{y}_n)$$

$$= \left(x_{n+1}'' + \sum_{j=2}^{n} x_j, T\left(x_n'' + \sum_{j=2}^{n} x_j\right)\right)$$

$$\in [(x, y) + V \times B_{n-1}] \bigcap \Gamma(T)$$

□

Some of the consequences of the closed graph theorem are downright weird. For example:

Corollary 4.38. *Suppose X and Y are Hausdorff locally convex spaces, and suppose X is barreled and Y is a Fréchet space. Suppose T is a linear transformation from X to Y, and suppose Z is a Hausdorff space and $F : Y \rightarrow Z$ is a one-to-one continuous function for which $F \circ T$ is continuous. Then T is continuous.*

Proof. $\Gamma(T) = (\mathrm{id} \times F)^{-1}(\Gamma(F \circ T))$ is closed.   □

What makes this useful is that continuity can be deduced using a coarser topology [the topology from Z induced on $F(Y)$, then pulled back to Y]. When the topology is coarser, it is easier for a function to be continuous (fewer open sets with inverse images that must be open) but harder to have closed graph (the topology on $X \times Y$ is coarser). The closed graph theorem actually provides continuity in a fine topology on Y, which after all is a Mackey space (Theorem 4.5 plus Corollary 4.9).

Example 5. Suppose Z is a Hausdorff topological vector space, and X is a subspace. Call X "Fréchetable" if X can be equipped with a Fréchet space structure for which the inclusion $X \hookrightarrow Z$ is continuous. If so, that Fréchet space structure is unique. (Let $Y = X$ as sets in Corollary 4.38, but with possibly different topologies on X and Y. Then reverse their roles.) For example, $AC_0[0, 1]$, the space of absolutely continuous functions f on $[0, 1]$ for which $f(0) = 0$, sits inside $C([0, 1])$, and is a Banach space, with the norm being total variation. Total variation? Where did that come from? Simply from how it sits as a subspace of $C([0, 1])$, $AC_0[0, 1]$ somehow "invents" total variation! Happy Halloween!

We close with a result commonly stated for Banach spaces. In that context, it is an easy consequence of either the closed graph theorem (which is why it is here) or the uniform boundedness principle. In fact, very little is needed. What really makes it work is the fact that locally convex spaces are locally weakly closed.

Proposition 4.39. *Suppose X and Y are Hausdorff locally convex spaces, and suppose X is infrabarreled. Suppose $T : X \to Y$ is a linear transformation for which $f \circ T \in X^*$ whenever $f \in Y^*$ Then T is continuous.*

Proof. First of all, if $f \in Y^*$, then $x \in \{f\}_\circ \Leftrightarrow |f(x)| \leq 1$, so

$$T^{-1}(\{f\}_\circ) = T^{-1}(f^{-1}(\{z : |z| \leq 1\}))$$

$$= (f \circ T)^{-1}(\{z : |z| \leq 1\})$$

is a neighborhood of 0 in X, and

$$T^{-1}(\{f_1, \ldots, f_n\}_\circ) = \bigcap_{j=1}^{n} (f_j \circ T)^{-1}(\{z : |z| \leq 1\})$$

is a neighborhood of 0 in X. Letting Y_w denote Y with the weak topology, $T \in \mathscr{L}_c(X, Y_w)$ by Proposition 1.26(a).

Let B be a barrel neighborhood of 0 in Y; then $T^{-1}(B)$ is convex, balanced, and absorbent. [As before, $T^{-1}(B)$ absorbs x because B absorbs $T(x)$.] But $T^{-1}(B)$ is also closed, since B is weakly closed. Now suppose A is bounded in X. Then every $f \circ T$ is bounded on A when $f \in Y^*$, so every $f \in Y^*$ is bounded on $T(A)$. Hence $T(A)$ is bounded in Y by Corollary 3.31. Hence $T(A) \subset cB$ for some c, so $A \subset T^{-1}(T(A)) \subset T^{-1}(cB) = cT^{-1}(B)$. That is, $T^{-1}(B)$ absorbs A. Since A was arbitrary, $T^{-1}(B)$ is a neighborhood of 0 since X is infrabarreled. Since B was an arbitrary barrel neighborhood of 0 in Y, T is continuous by Proposition 1.26(a).

$\square$

A final comment. The closed graph theorem will be proven in a different context in Chap. 5, where the assumption that Y is a Fréchet space will be replaced by a different, more general, assumption. The proof will be based on Proposition 4.39 above; the fact that $\Gamma(T)$ is closed basically will be used to get that $(X \times Y)/\Gamma(T)$

is a Hausdorff space. However, the fact that the assumption on Y is more general is deep, usually depending on Krein–Smulian I in Chap. 6. (An alternate, independent approach, appears in Exercise 20 of Chap. 5.) The proof given in this section is a lot cheaper to get.

Exercises

1. Suppose X is a Hausdorff locally convex space, and B is a nonempty closed, convex, balanced subset of X.

   (a) Show that B is a barrel if, and only if, $B°$ is weak-* bounded in X^*.
   (b) Show that B absorbs all bounded sets in X if, and only if, $B°$ is strongly bounded in X^*.

2. Suppose X is a Hausdorff locally convex space, and suppose X is infrabarreled. Show that a strongly bounded, weak-* closed subset of X^* is weak-* compact.

3. (Partial converse to #2) Suppose X is a Hausdorff locally convex space, and suppose that any strongly bounded, weak-* closed subset of X^* is weak-* compact. Show that the Mackey topology on X is infrabarreled.

4. Suppose X and Y are Hausdorff locally convex spaces over the same field $\mathbb{F}$ ($\mathbb{R}$ or $\mathbb{C}$), and suppose $B : X \times Y \to \mathbb{F}$ is a separately continuous bilinear form. Show that if X is barreled, then the map

   $$X \to Y^* = \text{ strong dual of } Y$$

   $$x \mapsto B(x, ?)$$

   is continuous. Use this to prove the Hellinger–Toeplitz theorem:

   If X is barreled, and $T : X \to X^*$ is a function such that for all $x, y \in X : T(x)(y) = T(y)(x)$, then T is a continuous linear transformation.

5. Suppose X is a Hausdorff locally convex space over the base field $\mathbb{F}$ ($\mathbb{R}$ or $\mathbb{C}$), and suppose the evaluation map

   $$X^* \times X \to \mathbb{F}$$

   $$(f, x) \mapsto f(x)$$

   is jointly continuous. Show that the topology of X can be given by a norm. *Hint:* $\{z \in \mathbb{F} : |z| \leq 1\}$ is a perfectly good neighborhood of 0 in $\mathbb{F}$, and it works well with polars.

6. Using Exercise 5, show that if the dual of a Fréchet space is another Fréchet space, then the topology of the original space can be given by a norm.

7. Suppose X is a Hausdorff locally convex space. Show that X is quasi-complete if and only if closed, bounded sets are complete.

8. Suppose X is a Hausdorff locally convex space, with topology τ_0 and dual space X^*. Let τ_M denote the Mackey topology on X, and suppose τ_1 is a locally convex topology on X for which $\tau_M \supset \tau_1 \supset \tau_0$. Finally, suppose A is a subset of X which is complete in the topology τ_0. Show that A is complete in the topology τ_1.

   Hint: The topology τ_1 is locally weakly closed, hence is locally τ_0-closed. (Use barrel neighborhoods.) Now look closely at the completeness parts of the proofs of Proposition 3.30 and Theorem 4.20.

9. Suppose X is a Hausdorff locally convex space, with topology τ_0 and dual space X^*. Let τ_M denote the Mackey topology on X, and suppose τ_1 is a locally convex topology on X for which $\tau_M \supset \tau_1 \supset \tau_0$. Using the preceding two problems, prove:

   (a) If (X, τ_0) is complete, the (X, τ_1) is complete.
   (b) If (X, τ_0) is quasi-complete, then (X, τ_1) is quasi-complete.

10. Suppose X and Y are Hausdorff locally convex spaces, with X being infrabarreled and Y being sequentially complete. Show that $\mathscr{L}_c(X, Y)$ is sequentially complete in the topology of bounded convergence.

11. Suppose X is a Hausdorff locally convex space. Show that X^* is complete in the weak-* topology if, and only if, $X^* = X'$.

12. Suppose X is a locally convex space, and $A \subset X$. Show that if $A \subset B \subset (A^\circ)_\circ$, then $A^\circ = B^\circ$. Do this without choosing elements from either side; use set containments and the bipolar theorem.

13. Suppose Y is a topological space and X is a dense subspace. Show that the inclusion map $X \hookrightarrow Y$ is nearly open.

14. Suppose X is a locally compact Hausdorff space, and Y is a Hausdorff space. Show that if $f : X \to Y$ is continuous and nearly open, then f is an open map.

15. Set

$$X = \bigcup_{n=1}^{\infty} [0, 1] \cdot e^{i\pi/n} \text{ in } \mathbb{C} \approx \mathbb{R}^2, \text{ and}$$

$$Y = X^- = X \cup [0, 1]$$

equip Y with the usual Euclidean metric, and X with the "Washington metric," where

$$d(z, w) = \begin{cases} |z - w| \text{ if } z \text{ and } w \text{ are colinear} \\ |z| + |w| \text{ if not.} \end{cases}$$

Show that $\mathbb{R}^2$, with the Washington metric, is complete, and show that X is closed in this metric. Finally, show that $X \hookrightarrow Y$ is continuous and nearly open, but not open.

16. Suppose X and Y are two Hausdorff locally convex spaces.

   (a) Show that if X and Y are complete, then so is $X \times Y$.
   (b) Show that if X and Y are first countable, then so is $X \times Y$.
       (Moral: If X and Y are Fréchet spaces, then so is $X \times Y$.)

17. Suppose X is a Fréchet space and Y is barreled, and suppose $T : X \to Y$ has dense range but is not onto. Show that the algebraic dimension of $Y/T(X)$ is infinite. *Hint:* If not, choose a finite-dimensional complementary subspace Z, and consider $Z \times X \to Y$, $(z, x) \mapsto z + T(x)$. Use the open mapping theorem. (See Exercise 16.)

18. (Converse to Theorem 4.12) Suppose Y is a Hausdorff locally convex space, and suppose $\mathscr{L}_b(Y, Z) = \mathscr{L}_c(Y, Z)$ whenever Z is a normed space. Show that Y is bornological. *Suggestion:* If C is a convex, balanced subset of Y that absorbs all bounded sets, let p_C denote the associated Minkowski functional. Let X denote the vector space Y topologized with the seminorm p_C, and apply Chap. 3, Exercise 9.

19. (See previous exercise.) Suppose X is a Hausdorff locally convex space. Show that X is bornological provided X is infrabarreled and every bounded linear functional on X is continuous.

The next six problems are concerned with the closed graph theorem. The first two are very similar. The trick for the first four (as is often the case) is to come up with the appropriate spaces and maps.

20. Let μ denote Lebesgue measure on $\mathbb{R}$. Suppose $g : \mathbb{R} \to \mathbb{R}$ is a measurable function with the property that for all measurable $f : \mathbb{R} \to \mathbb{R}$:

$$\int |f|^2 d\mu < \infty \Rightarrow \int |fg| d\mu < \infty.$$

Show that $\int |g|^2 d\mu < \infty$.

21. Suppose $(X, \mathscr{B})$ is a measurable space, and suppose μ and ν are two finite measures on $(X, \mathscr{B})$ having the same sets of measure zero. Finally, suppose $1 < p < \infty$ and $1 = \frac{1}{p} + \frac{1}{q}$. Show that $L^p(\mu) \subset L^1(\nu)$ if, and only if, $[d\nu/d\mu] \in L^q(\mu)$. Here, $[d\nu/d\mu]$ denotes the function class of the Radon–Nikodym derivative.

22. (Weak integrals) Suppose $(X, \mathscr{B}, \mu)$ is a measure space, and $\mathscr{H}$ is a Hilbert space. Suppose $f : X \to \mathscr{H}$ is a function with the property that for all $v \in \mathscr{H}$, $x \mapsto (f(x), v)$ is an integrable function. Show that there exists a vector $I(f) \in \mathscr{H}$, called the *weak integral* of f, for which

$$(I(f), v) = \int_X (f(x), v) d\mu(x).$$

Hint: If $\mathscr{H}$ is a Hilbert space over $\mathbb{C}$, look at $(v, f(x))$. Do you see why?

23. Suppose $\langle b_n \rangle$ is a sequence of real numbers such that for all real sequences $\langle a_n \rangle$:

$$\lim_{n \to \infty} a_n = 0 \Rightarrow \sum_{n=1}^{\infty} |a_n b_n| < \infty.$$

Show that $\sum |b_n| < \infty$.

24. Derive Corollary 4.36 from the closed graph theorem.

25. Suppose X is an LF-space and Y is a Fréchet space. Show that a linear map $T : X \to Y$ is continuous provided that its graph is sequentially closed.

Note: When doing Exercise 22, you probably had to use sequences. Lebesgue integrals work well with sequences, but not with nets, since measures are only countably additive.

The next three problems are purely topological, and will be used in the last three problems.

26. Suppose X is a topological space, and Y is a subspace. Suppose $A \subset Y$. Show that if A is nowhere dense in Y, then A is nowhere dense in X. (Recall: $A^- \cap Y$ is the closure of A in Y.)

27. Suppose X is a second category topological space. Show that X is not a countable union of first category subspaces.

28. Suppose X is a second category topological space, and suppose $X = \bigcup X_n$, where $X_1 \subset X_2 \subset \cdots$. Show that there exists N such that X_n is second category for $n \geq N$.

29. Show that an infinite-dimensional Fréchet space cannot have countable algebraic dimension, that is cannot have a countable Hamel basis.

30. (Bourbaki) Suppose X is an infinite dimensional Fréchet space. Let B_1 denote a countably infinite, linearly independent subset of X. Let Y denote the closure of the linear span of B_1, and extend B_1 to a Hamel basis B_2 of Y. B_2 is uncountable by Exercise 29. Finally, extend B_2 to a Hamel basis B_3 of X. Let $C = \{v_1, v_2, v_3, \cdots\}$ be a countably infinite subset of $B_2 - B_1$. Set

$$X_n = \text{algebraic linear span of } (B_3 - \{v_n, v_{n+1}, \ldots\}).$$

Show that $X = \bigcup X_n$, $X_1 \subsetneq X_2 \subsetneq X_3 \cdots$, and all X_n are dense in X. Hence by Exercise 28, X_n is second category beyond some point, and so is barreled. (Theorem 4.5.)

31. Suppose $Y = \cup Y_n$ is an LF-space, X is a Hausdorff locally convex space, and $T : X \to Y$ is a linear transformation with a graph that is closed in $X \times Y$.

   (a) Suppose X is a Fréchet space. Show that T is continuous. *Suggestion:* Use Exercise 19 from Chap. 1, with $\tilde{H} = Y_n$ for n sufficiently large: "Sufficiently large" is obtained from Exercise 28, with $X_n = T^{-1}(Y_n)$. Theorem 4.5 and Baire category also arise here, as does Theorem 4.37.

   (b) Suppose X is an LF-space. Show that T is continuous.

Chapter 5
Dual Spaces

5.1 Adjoints

Perhaps a better title for this chapter would be "Duality," but this has a special meaning in functional analysis: an abstraction of the notion of a space X and its dual X^* into a pairing (X, Y), where X is a vector space and Y is a space of linear functionals on X. The subject has its uses; the argument in Proposition 5.38 of Sect. 5.7 is based on such concepts. However, it leads away from the more practical functional analysis that is the subject of this book. Typically, one has some fairly specific spaces in mind, which then dictate the dual structure.

Probably the most fundamental concept here that is *not* normally discussed in beginning graduate real analysis is the notion of an adjoint map (although the underlying idea often does appear in beginning graduate algebra!).

Definition 5.1. Suppose X and Y are locally convex spaces, and $T : X \to Y$ is a continuous linear map. The **adjoint** of T, denoted by T^*, is the map from Y^* to X^* defined by $T^*(f) = f \circ T$. That is,

$$T^* : Y^* \to X^*$$

$$f \mapsto f \circ T$$

$$\forall x \in X, \forall f \in Y^* \;\; : \;\; [T^*(f)](x) = f[T(x)].$$

The preceding is a bit pedantic because the concept is slippery, appearing simultaneously trivial and mind-bending, particularly if one is interested in establishing results about T^*. Consider the following.

Theorem 5.2. *Suppose X and Y are locally convex spaces, and $T : X \to Y$ is a continuous linear map. Letting X^* and Y^* denote their strong dual spaces, $T^* : Y^* \to X^*$ is a continuous linear map. Also, if $A \subset X$, then $T(A)^\circ = (T^*)^{-1}(A^\circ)$. Finally, T^* is continuous when X^* and Y^* are equipped with their weak-* topologies.*

M.S. Osborne, *Locally Convex Spaces*, Graduate Texts in Mathematics 269,   123
DOI 10.1007/978-3-319-02045-7_5, © Springer International Publishing Switzerland 2014

Proof. Letting $\mathbb{F}$ denote the base field, if $c \in \mathbb{F}$, $x \in X$, and $f, g \in Y^*$, then

$$T^*(cf)(x) = (cf)(T(x)) = cf(T(x))$$
$$= c[T^*(f)(x)] = [cT^*(f)](x), \text{ and}$$
$$T^*(f+g)(x) = (f+g)(T(x)) = f(T(x)) + g(T(x))$$
$$= T^*(f)(x) + T^*(g)(x) = [T^*(f) + T^*(g)](x),$$

so T^* is linear.

Next, suppose $A \subset X$. Then for all $f \in Y^*$:

$$f \in T(A)^\circ \Leftrightarrow \forall x \in A : |f(T(x))| \leq 1$$
$$\Leftrightarrow \forall x \in A : |T^*(f)(x)| \leq 1$$
$$\Leftrightarrow T^*(f) \in A^\circ \Leftrightarrow f \in (T^*)^{-1}(A^\circ).$$

In particular, if F is finite in X, then $(T^*)^{-1}(F^\circ) = T(F)^\circ$, so T^* is continuous when X^* and Y^* have their weak-* topologies.

Finally, T^* is strongly continuous since T is bounded: If A is bounded in X, so that A° is a typical neighborhood of 0 in X^*, then $(T^*)^{-1}(A^\circ) = (T(A))^\circ$ is a strong neighborhood of 0 in Y^*.   □

Now for the subtleties. The fact that T^* is linear is not a surprise, but note that it has *nothing to do* with the fact that T is linear. It simply follows from how $\mathbb{F}$-valued functions are added together or multiplied by scalars. Also, the continuity proof made no use of the fact that T was continuous, only that T was bounded. What the (stronger) continuity condition does is guarantee that T^* takes values in X^*. If T were only bounded, then $T^*(f)$ would be a bounded (but not necessarily continuous) linear functional.

There are two more results to establish.

Proposition 5.3. *Suppose X, Y, and Z are locally convex spaces, and $T \in \mathscr{L}_c(X, Y)$ and $S \in \mathscr{L}_c(Y, Z)$. Then $ST = S \circ T \in \mathscr{L}_c(X, Z)$, and $(ST)^* = T^*S^*$.*

Proof. If $x \in X$ and $f \in Z^*$, then

$$(ST)^*(f)(x) = f(S \circ T(x)) = f(S[T(x)])$$
$$= S^*(f)[T(x)] = T^*[S^*(f)](x).$$ □

Proposition 5.4. *Suppose X and Y are locally convex spaces. Equip $\mathscr{L}_c(X, Y)$ and $\mathscr{L}_c(Y^*, X^*)$ with their topologies of bounded convergence. Then $T \mapsto T^*$ is a linear map from $\mathscr{L}_c(X, Y)$ to $\mathscr{L}_c(Y^*, X^*)$, which is continuous when Y is infrabarreled.*

Proof. If $S, T \in \mathscr{L}_c(X, Y)$, then for all $f \in Y^*$, $x \in X$, and scalar c:

$$(cT)^*(f)(x) = f(cT(x)) = cf(T(x))$$
$$= c[T^*(f)(x)] = [cT^*](f)(x) \text{ and}$$
$$(S + T)^*(f)(x) = f[(S + T)(x)] = f[S(x) + T(x)]$$
$$= f[S(x)] + f[T(x)] = S^*(f)(x) + T^*(f)(x)$$
$$= [S^*(f) + T^*(f)](x) = [S^* + T^*](f)(x).$$

Notice! The linearity of T plays no role here, but the linearity of f does.

Now suppose Y is infrabarreled. A typical neighborhood of 0 in $\mathscr{L}_c(Y^*, X^*)$ has the form $N(E, A^\circ)$, where E is strongly bounded in Y^* and A is bounded in X (so that A° is a typical member of the neighborhood base at 0 defining the strong topology). Then E is equicontinuous by Theorem 4.16(b), so

$$E_\circ = \{y \in Y : |f(y)| \le 1 \text{ for all } f \in E\}$$
$$= \bigcap_{f \in E} f^{-1}(\{c \in \mathbb{F} : |c| \le 1\})$$

is a neighborhood of 0 in Y. But now,

$$T \in N(A, E_\circ) \Leftrightarrow \forall x \in A : T(x) \in E_\circ$$
$$\Leftrightarrow \forall x \in A, \forall f \in E : |f[T(x)]| \le 1$$
$$\Leftrightarrow \forall f \in E : f \in T(A)^\circ$$
$$\Leftrightarrow E \subset T(A)^\circ = (T^*)^{-1}(A^\circ) \qquad \text{(Thm. 5.2)}$$
$$\Leftrightarrow T^*(E) \subset A^\circ \Leftrightarrow T^* \in N(E, A^\circ).$$

Hence the map $T \mapsto T^*$ maps $N(A, E_\circ)$ into $N(E, A^\circ)$. $\qquad\square$

In the preceding, note that if Y is infrabarreled, then two things happen. In the first place, looking more closely at Theorem 4.16, the strongly bounded sets in Y^* are *precisely* the equicontinuous sets, so if U is a barrel neighborhood of 0 in Y (so that $U = (U^\circ)_\circ$), then $T \mapsto T^*$ maps $N(A, U)$ into $N(U^\circ, A^\circ)$. Also, $T \mapsto T^*$ has trivial kernel, since Y is Hausdorff (infrabarreled was defined that way!): If $T^* = 0$, then for all $f \in Y^*$ and $x \in X$: $f(T(x)) = T^*(f)(x) = 0$, so $T(x) = 0$ by Corollary 3.17 (Y^* separates points). Since x was arbitrary, $T = 0$. Thus, $T \mapsto T^*$ is a homeomorphism of $\mathscr{L}_c(X, Y)$ with its image in $\mathscr{L}_c(Y^*, X^*)$ when Y is infrabarreled. The following are left to the exercises: $T \mapsto T^*$ is an isometry when X and Y are normed, and $T \mapsto T^*$ is bounded whether Y is infrabarreled or not.

There are some more things about T^* that need to be said, but they depend on some structural results, to be discussed in the next section. These structural results make use of what has just been established concerning adjoints.

5.2   Subspaces and Quotients

Suppose X is a locally convex space, and M is a subspace. Then M is a locally convex space in its own right, and it should be possible to say something intelligent about its dual space. The same holds for the quotient space X/M. Buried in there should be a calculation of the dual space of a product space. In fact, all this is true. One new (sort of—see below) idea is needed.

Definition 5.5. Suppose X is a locally convex space, and M is a subspace. Define $M^\perp \subset X^*$ as follows:

$$M^\perp = \{f \in X^* : f|_M \equiv 0\}.$$

Similarly, if N is a subspace of X^*, define $N_\perp \subset X$ as follows:

$$N_\perp = \{x \in X : f(x) = 0 \text{ for all } f \in N\}.$$

In a sense, the preceding is not needed, since for a sub*space* M of X and a sub*space* N of X^*,

$$M^\perp = M^\circ \text{ and } N_\perp = N_\circ.$$

The reason is that if $f \in X^*$, then $f(M)$ is either all of the base field (in which case $f \notin M^\circ$ and $f \notin M^\perp$) or is 0 alone (in which case $f \in M^\circ$ and $f \in M^\perp$). A similar argument works for N. Consequently, the bipolar theorem already establishes, for example, that $(M^\perp)_\perp = M$ when M is closed. The main reason for introducing the "$\perp$" notation is for emphasis. (Not all follow this: Royden [30], for example, uses the "$\circ$" notation, although he only uses it for subspaces.)

Theorem 5.6. *Suppose X is a locally convex space with strong dual X^*, and M is a subspace of X. Equip M with the induced topology, and let M^* denote its strong dual. Let $\iota : M \hookrightarrow X$ denote the inclusion map. Then*

(a) Algebraically, $\iota^* : X^* \to M^*$ is the restriction mapping $f \mapsto f|_M$.
(b) $\iota^* : X^* \to M^*$ is onto, with kernel $M^\perp$.
(c) The induced algebraic isomorphism $(\iota^*)_0 : X^*/M^\perp \to M^*$ is a continuous bijection.
(d) If M is infrabarreled, then $(\iota^*)_0^{-1} : M^* \to X^*/M^\perp$ is bounded.

Proof. (a) For all $x \in M$ and $f \in X^*$: $f(\iota(x)) = f(x) = f|_M(x)$. (Yes, that is all there is to it!)
(b) ι^* is onto by Proposition 3.16, while trivially $f|_M \equiv 0 \Leftrightarrow f \in M^\perp$.
(c) ι^* is continuous by Theorem 5.2, so $(\iota^*)_0$ is continuous by Theorem 1.23(c).
(d) Suppose E is bounded in M^*. Then E is equicontinuous by Theorem 4.16(b), since the strong dual has the topology of bounded convergence. In particular,

$$E_\circ = \{x \in M : |f(x)| \leq 1 \text{ for all } f \in E\}$$

is a neighborhood of 0 in M, so there exists a convex, balanced neighborhood B of 0 in X for which $B \cap M \subset E_\circ$. Let p_B denote the support functional for B. Since B is absorbent, p_B is defined on all of X. Since B is convex and balanced, p_B is a seminorm (Theorem 3.7). There are now two cases to consider:

Base field $\mathbb{R}$: Then, for all $f \in E$:

$$x \in B \cap M \Rightarrow x \in E_\circ \Rightarrow f(x) \le |f(x)| \le 1.$$

Hence $f(x) \le p_{B \cap M}(x)$ for $x \in M$ by Proposition 3.8. But $p_{B \cap M} = p_B\big|_M$ by definition, so $f(x) \le p_B(x)$ for $x \in M$. The Hahn–Banach theorem now extends f to a continuous linear functional F on X for which $F(x) \le p_B(x)$. (p_B Is continuous since B is a neighborhood of 0 [Theorem 3.7].) Hence $F(x) \le 1$ for $x \in B$ by Proposition 3.8. But now $\pm F(x) = F(\pm x) \le 1$ for $x \in B$ since B is balanced, so $|F(x)| \le 1$ for $x \in B$. Hence $F \in B^\circ$.

Base field $\mathbb{C}$: Then, for all $f \in E$:

$$x \in B \cap M \Rightarrow x \in E_\circ \Rightarrow Re f(x) \le |f(x)| \le 1.$$

Hence (again) $Re f(x) \le p_B(x)$ for $x \in M$ by Proposition 3.8. Again, the Hahn-Banach theorem extends $Re f$ to a continuous linear functional $g : X \to \mathbb{R}$ for which $g(x) \le p_B(x)$. By Proposition 3.14, there is $F \in X^*$ for which $Re F = g$. Since $F\big|_M$ and f have the same real parts, they are equal (Proposition 3.14 again). Finally, since $Re F(x) = g(x) \le p_B(x)$ for $x \in X$, we have that $Re F(x) \le 1$ for $x \in B$ by Proposition 3.8. Hence $|F(x)| \le 1$ for $x \in B$ since B is balanced. That is, $F \in B^\circ$.

So, in either case, we have, for each $f \in E$, an extension F to X for which $F \in B^\circ$. But B° is by definition equicontinuous, hence is bounded (Theorem 4.16(a)). Let $\pi : X^* \to X^*/M^\perp$ denote the natural map. Since $\iota^*(F) = F\big|_M = f$, we have that $(\iota^*)_0(F + M^\perp) = f$, that is $(\iota^*)_0^{-1}(f) = F + M^\perp \in \pi(B^\circ)$. But π is continuous, so $\pi(B^\circ)$ is bounded in $X^*/M^\perp$. Since $\pi(B^\circ) \supset (\iota^*)_0^{-1}(E)$, $(\iota^*)_0^{-1}(E)$ is bounded. $\qquad\square$

Of course, if X is a Banach space, then all these spaces are normed, hence first countable, hence bornological, hence infrabarreled, so $(\iota^*)_0^{-1}$ is continuous for Banach spaces. This does not happen in general, and similar restrictions hold for duals of quotients.

Theorem 5.7. *Suppose X is a locally convex space with strong dual X^*, and M is a subspace of X. Equip X/M with the quotient topology, and let $(X/M)^*$ denote its strong dual. Let $\pi : X \to X/M$ denote the natural projection. Then:*

(a) *Algebraically, $\pi^* : (X/M)^* \to X^*$ is defined by $\pi^*(f)(x) = f(x + M)$.*
(b) *$\pi^* : (X/M)^* \to X^*$ is one-to-one, with image $M^\perp$.*
(c) *The induced algebraic isomorphism $(\pi^*)_0 : (X/M)^* \to M^\perp$ is a continuous bijection.*
(d) *If X is infrabarreled, then $(\pi^*)_0^{-1} : M^\perp \to (X/M)^*$ is bounded.*

Proof. (a) For all $x \in X$ and $f \in (X/M)^*$, by definition $\pi^*(f)(x) = f(\pi(x)) = f(x + M)$. (Again, that is all there is to it.)

(b) $\pi^*(f) = 0$ when $f(x + M) = 0$ for all $x \in X$, that is only when $f = 0$, so π^* is one-to-one. By the usual business for homomorphisms [and Theorem 1.23(c)], a continuous linear functional on X/M corresponds directly [via the formula in part (a)] to a continuous linear functional on X which vanishes on M—that is, to a member of $M^\perp$. Hence $M^\perp$ is the image of π^*.

(c) π^* is continuous by Theorem 5.2, so $(\pi^*)_0$ is continuous (Proposition 1.21).

(d) Suppose E is bounded in $M^\perp$. Then E is bounded in X^*, hence is equicontinuous by Theorem 4.16(b). Hence $E_\circ$ is a neighborhood of 0 in X. Hence $\pi(E_\circ)$ is a neighborhood of 0 in X/M by Proposition 1.26(b) since $\pi : X \to X/M$ is an open map (Theorem 1.23(b)). Thus, $(\pi(E_\circ))^\circ$ is equicontinuous, hence is bounded (Theorem 4.16(a)), in $(X/M)^*$. It therefore suffices to show that $(\pi^*)_0^{-1}(E) \subset (\pi(E_\circ))^\circ$. This is basically a matter of unraveling the notation.

Suppose $f \in (\pi^*)_0^{-1}(E)$. Then $\pi^*(f) = (\pi^*)_0(f) \in E$. Now if $x \in E_\circ$, so that $x + M = \pi(x) \in \pi(E_\circ)$, then $|\pi^*(f)(x)| \leq 1$ by definition of $E_\circ$. That is, by part (a): $|f(x + M)| \leq 1$. That is, $|f(x + M)| \leq 1$ whenever $x + M = \pi(x) \in \pi(E_\circ)$. Hence by definition, $f \in (\pi(E_\circ))^\circ$. $\square$

Finally, the situation for direct products is much more satisfactory, due to a category theoretic construction, known as a biproduct.

Lemma 5.8. *Suppose X_1, X_2, and Y are locally convex spaces, and suppose there are continuous linear maps*

$$\begin{cases} \iota_k : X_k \to Y, & k = 1, 2, \text{ and} \\ \pi_k : Y \to X_k, & k = 1, 2 \end{cases}$$

subject to

 (i) $\pi_k \iota_k = $ identity on X_k, and
 (ii) $\iota_1 \pi_1 + \iota_2 \pi_2 = $ identity on Y.

Then:

 (a) $\pi_2 \iota_1 : X_1 \to X_2$ is the zero map,
 (b) $\pi_1 \iota_2 : X_2 \to X_1$ is the zero map, and
 (c) $\pi = (\pi_1, \pi_2) : Y \to X_1 \times X_2$ is a topological isomorphism. Its inverse is given by $(x_1, x_2) \mapsto \iota_1(x_1) + \iota_2(x_2)$.

Proof. First of all, since $\pi_k \iota_k$ is the identity, it is a bijection, so π_k is onto and ι_k is one-to-one. But now

$$\iota_1 = (\iota_1 \pi_1 + \iota_2 \pi_2)\iota_1$$
$$= \iota_1(\pi_1 \iota_1) + \iota_2(\pi_2 \iota_1)$$
$$= \iota_1 + \iota_2(\pi_2 \iota_1) \qquad \text{(since } \pi_1 \iota_1 = id.)$$

Hence $\iota_2(\pi_2\iota_1) = 0$, so $\pi_2\iota_1 = 0$ since ι_2 is one-to-one. This proves part (a). Part (b) is immediate by exchange: $1 \leftrightarrow 2$.

For part (c), the idea is to show that if one defines $\iota : X_1 \times X_2 \to Y$ by $\iota(x_1, x_2) = \iota_1(x_1) + \iota_2(x_2)$, then both π and ι are continuous, and are inverse to each other. This will complete the proof.

Continuity of π : immediate from Theorem 1.18.

Continuity of ι : We have continuity of

$$(x_1, x_2) \mapsto x_1 \mapsto \iota_1(x_1) \text{ and}$$

$$(x_1, x_2) \mapsto x_2 \mapsto \iota_2(x_2), \text{ so}$$

$$(x_1, x_2) \mapsto (\iota_1(x_1), \iota_2(x_2)) \mapsto \iota_1(x_1) + \iota_2(x_2)$$

$$X_1 \times X_2 \to Y \times Y \xrightarrow{\text{sum}} Y$$

is continuous.

$\pi \circ \iota$ is the identity:

$$\pi \circ \iota\,((x_1, x_2)) = \pi(\iota_1(x_1) + \iota_2(x_2))$$

$$= (\pi_1(\iota_1(x_1) + \iota_2(x_2)), \pi_2(\iota_1(x_1) + \iota_2(x_2)))$$

$$= (\pi_1\iota_1(x_1) + \pi_1\iota_2(x_2), \pi_2\iota_1(x_1) + \pi_2\iota_2(x_2))$$

$$= (x_1 + 0, 0 + x_2)$$

$\iota \circ \pi$ is the identity:

$$\iota \circ \pi(y) = \iota(\pi_1(y), \pi_2(y))$$

$$= \iota_1\pi_1(y) + \iota_2\pi_2(y) = y.$$

$\square$

Theorem 5.9. *Suppose X_1 and X_2 are locally convex spaces. Then $(X_1 \times X_2)^*$ is topologically isomorphic to $X_1^* \times X_2^*$, where any $(f, g) \in X_1^* \times X_2^*$ corresponds to the linear functional $(x_1, x_2) \mapsto f(x_1) + g(x_2)$.*

Proof. We have natural maps

$$\iota_k : X_k \to X_1 \times X_2$$

given by $\iota_1(x_1) = (x_1, 0)$ and $\iota_2(x_2) = (0, x_2)$. We also have natural projections $\pi_k : (X_1 \times X_2) \to X_k$, and these satisfy the equations

$$\pi_1\iota_1 = \text{identity on } X_1,$$

$$\pi_2\iota_2 = \text{identity on } X_2, \text{ and}$$

$$\iota_1\pi_1 + \iota_2\pi_2 = \text{identity on } X_1 \times X_2.$$

Taking adjoints:

$$\iota_1^* \pi_1^* = \text{identity on } X_1^*,$$

$$\iota_2^* \pi_2^* = \text{identity on } X_2^*, \text{ and}$$

$$\pi_1^* \iota_1^* + \pi_2^* \iota_2^* = \text{identity on } (X_1 \times X_2)^*.$$

Hence (with pi and iota reversed), $(X_1 \times X_2)^* \approx X_1^* \times X_2^*$ by Lemma 5.8.   □

In case you are curious: Yes, the dual of a product was deferred until now so that the adjoint maps above would be available to trivialize matters. Also, some of the preceding results generalize to spaces of linear transformations. There is even a generalization of Theorem 5.9 for infinite products: The dual of an infinite product is the *direct sum* of their duals—but with the box topology. This may come as a surprise: After all, the *algebraic* dual of a direct sum is a direct product, while the algebraic dual of a direct product is HUGE: It contains functionals with no simple coordinate formula, a fact easily deduced from the fact that there must exist linear functionals on the direct product which vanish on the direct sum. However, for topological duals, the product topology is so coarse that the dual of a product is severely constrained. The basic generalizations discussed here appear in the exercises: none are particularly difficult, given the methods discussed in this section.

5.3   The Second Dual

Any locally convex space X has a dual space X^*, and X^* is most naturally topologized using the strong topology. Consequently, X^* becomes a locally convex space, and so has a dual of its own, X^{**}. In textbooks, X^{**} is commonly called the **bidual** when the book discusses general spaces, and is often called the **second dual** when the subject is Banach spaces. The discussion here will sometimes make reference to X^{***} as well, and "third dual" is slightly more descriptive than "tridual." Consequently, in what follows, X^{**} will be referred to as the second dual.

One more thing. In order to properly discuss the natural map from X to its second dual, it will be necessary to assume that X is Hausdorff, so for the remainder of this section *all locally convex spaces under discussion will be assumed to be Hausdorff*.

Suppose X is a locally convex space, and $x \in X$. Then the evaluation map $f \mapsto f(x)$ is weak-* continuous on X^*, hence is strongly continuous. It is also linear; all this came up in Sect. 3.6, Corollary 3.23. As such, we may associate with x an element $J_X(x) \in X^{**}$ defined by the equation

$$J_X(x)(f) = f(x).$$

(By the way, there is no standard notation for the map J_X. The letter J is used by Yosida [41], and works well here.)

Theorem 5.10. *Suppose X is a Hausdorff locally convex space, and $J_X : X \to X^{**}$ is the natural map from X to its second dual. Then:*

(a) J_X is linear and one-to-one.
(b) If $E \subset X^*$, then $J_X(E_o) = E^\circ \cap J_X(X)$ and $J_X^{-1}(E^\circ) = E_o$.
(c) J_X is bounded and $J_X^{-1} : J_X(X) \to X$ is continuous.
(d) J_X is continuous if, and only if, X is infrabarreled.

Proof. (a) J_X is linear because the functionals to which it is applied are linear:

$$J_X(cx)(f) = f(cx) = cf(x) = cJ_X(x)(f) \text{ and}$$

$$J_X(x+y)(f) = f(x+y) = f(x) + f(y) = J_X(x)(f) + J_X(y)(f).$$

J_X is one-to-one because its kernel is zero: If $J_X(x) = 0$, then $f(x) = 0$ for all $f \in X^*$, giving that $x = 0$ (Corollary 3.17).

(b) Suppose $\varphi \in X^{**}$. Then $\varphi \in E^\circ$ if, and only if, $|\varphi(f)| \leq 1$ for all $f \in E$. If $\varphi = J_X(x)$, then $\varphi(f) = f(x)$, so

$$\varphi \in E^\circ \Leftrightarrow |f(x)| \leq 1 \text{ for all } f \in E \Leftrightarrow x \in E_o.$$

This proves both parts.

(c) If B is bounded in X, then $E = B^\circ$ is a neighborhood of 0 in X^*, so that E° is equicontinuous in X^{**}, hence is strongly bounded in X^{**} [Theorem 4.16(a)]. But $B \subset (B^\circ)_o = E_o$, so $J_X(B) \subset J_X(E_o) \subset E^\circ$, which is bounded. If U is barrel neighborhood of zero in X, then $(J_X^{-1})^{-1}(U) = J_X(U)$ in $J_X(X)$, and since $U = (U^\circ)_o$ by the Bipolar Theorem, setting $E = U^\circ$ gives

$$J_X(U) = J_X(E_o) = E^\circ \bigcap J_X(X).$$

by part (b). Now U is a neighborhood of 0, so E is equicontinuous, hence is strongly bounded in X^* [Theorem 4.16(a) again]. Hence E° is a neighborhood of 0 in X^{**}.

(d) Suppose X is infrabarreled, and E is strongly bounded in X^*. Then E° is a typical base neighborhood of 0 in X^{**}, and $J_X^{-1}(E^\circ) = E_o$ by part (b). But E is equicontinuous by Theorem 4.16(b), so that E_o is a neighborhood of zero in X. Finally, suppose J_X is continuous. Suppose B is a barrel in X which absorbs all bounded sets A. Set $E = B^\circ$; then $E_o = B$ by the bipolar theorem. Furthermore, E_o absorbs A whenever A is bounded, so A° absorbs E whenever A is bounded [Theorem 3.20(e)]. That is, E is strongly bounded, so E° is a neighborhood of zero in X^{**}. Hence $B = E_o = J_X^{-1}(E^\circ)$ is a neighborhood of zero in X. $\square$

Definition 5.11. Suppose X is a locally convex space. Then X is **semireflexive** if $J_X : X \to X^{**}$ is onto. X is **reflexive** if $J_X : X \to X^{**}$ is a topological isomorphism.

Of course, in view of Theorem 5.10, a space is reflexive if, and only if, it is semireflexive and infrabarreled. Somewhat more can be said, though. This requires a few more results.

Proposition 5.12. *Suppose X is a semireflexive, Hausdorff locally convex space. Then X^* is barreled.*

Proof. Suppose E is a barrel in X^*, so that E is strongly closed, convex, balanced, and absorbent. The fact that $J_X : X \to X^{**}$ is bijective says that the weak and weak-* topologies on X^* coincide, and E is weakly closed (Theorem 3.29), hence is weak-* closed. Hence $E = (E_\circ)^\circ$ by the bipolar theorem. Set $A = E_\circ$, so that $E = A^\circ$.

If $F \subset X^*$, with F finite, then E absorbs F since E is a barrel. That is, A° absorbs F, so $F_\circ$ absorbs A (Theorem 3.20(e)). That is, A is bounded in the weak topology on X, and so is originally bounded (Corollary 3.31). Hence $E = A^\circ$ is a standard base neighborhood of zero in X^*.   □

Before our next result, we need a pure (but weird) computation.

Lemma 5.13. *Suppose X is an infrabarreled, Hausdorff locally convex space, so that $J_X : X \to X^{**}$ has a continuous adjoint $J_X^* : X^{***} \to X^*$. Then $J_X^* \circ J_{X^*}$ is the identity map on X^*.*

Proof. Suppose $f \in X^*$, and $J_{X^*}(f) = \Phi$. If $x \in X$, then set $\psi = J_X(x)$. By definition:

$$J_X^*(\Phi)(x) = \Phi(J_X(x)) = \Phi(\psi)$$
$$= J_{X^*}(f)(\psi) = \psi(f)$$
$$= J_X(x)(f) = f(x).$$

Since x was arbitrary, $f = J_X^*(\Phi) = J_X^* \circ J_{X^*}(f)$.   □

Proposition 5.14. *Suppose X is a reflexive Hausdorff locally convex space. Then X^* is also reflexive.*

Proof. By assumption $J_X : X \to X^{**}$ is a topological isomorphism with inverse $J_X^{-1} : X^{**} \to X$, so $J_X^* : X^{***} \to X^*$ is a topological isomorphism with inverse $(J_X^{-1})^* : X^* \to X^{***}$. But Lemma 5.13 computes J_{X^*} as a right inverse for J_X^*, so J_{X^*} is $(J_X^{-1})^*$, a topological isomorphism.   □

Corollary 5.15. *Suppose X is a semireflexive Hausdorff locally convex space. Then the following are equivalent:*

 (i) X is reflexive.
 (ii) X is infrabarreled.
 (iii) X is barreled.

Proof. (iii) $\Rightarrow$ (ii) trivially, while (ii) $\Rightarrow$ (i) by Theorem 5.10. Suppose (i): that is, suppose X is reflexive. Then X^* is reflexive by Proposition 5.14, so X^{**} is barreled by Proposition 5.12. Hence X is barreled since X is topologically isomorphic to X^{**} by assumption. $\qquad\square$

For Banach spaces, Proposition 5.14 is an if-and-only-if. Not so in general; think of using a dense subspace of a reflexive space as the original space. However, some things can be said. Here is a start.

Proposition 5.16. *Suppose X is a Hausdorff locally convex space. Then X is semireflexive if, and only if, bounded, closed, convex subsets of X are weakly compact.*

Proof. First, suppose X is semireflexive, and B is bounded, closed, and convex in X. Then $B \subset (B°)_\circ$, and $(B°)_\circ$ is weakly closed, convex, bounded, and balanced. Set $E = B°$; then E is a strong neighborhood of zero in X^*, so $E°$ is weak-* compact in X^{**} by Theorem 3.26. Since J_X is a bijection of X with X^{**}, Theorem 5.10(b) (applied to finite sets) gives that J_X^{-1} is an isomorphism of $(X^{**}$, weak-* topology) with $(X$, weak topology). Since $J_X^{-1}(E°) = E_\circ \supset B$, and B is weakly closed: B is weakly compact.

Finally, suppose every closed, bounded, convex subset of X is weakly compact. The strong topology on X^* consists of taking the polars of bounded sets B, and $B° = ((B°)_\circ)°$ by the bipolar theorem. But $(B°)_\circ$ is weakly closed, convex, and bounded, so it is weakly compact. Thus, in fact, the strong topology on X^* is produced by taking polars of all weakly compact, convex subsets of X (since weakly compact $\Rightarrow$ weakly bounded $\Rightarrow$ bounded). This just gives the Mackey topology for X^* under the restriction that X [more literally, $J_X(X)$] is its dual space. Hence the dual of X^* is $J_X(X)$ by Proposition 3.27. $\qquad\square$

Corollary 5.17. *Suppose X is a Hausdorff locally convex space, and suppose X^* is semireflexive. Then X is infrabarreled if, and only if, X is a Mackey space.*

Proof. Infrabarreled spaces are Mackey spaces by Corollary 4.9, so suppose X is a Mackey space and X^* is semireflexive. If D is strongly bounded in X^*, then $E = (D°)_\circ$ is also strongly bounded, as well as weakly closed, convex, and balanced, so E is weakly compact in X^* by Proposition 5.16. Hence E is weak-* compact in the (coarser) weak-* topology on X^*, so $E_\circ$ is a Mackey neighborhood of zero in X. But $J_X^{-1}(E°) = E_\circ$, and $E° = D°$ by the bipolar theorem, so since $D°$ is a typical strong neighborhood of zero in X^{**}, J_X is continuous. Hence X is infrabarreled by Theorem 5.10(d). $\qquad\square$

The next result gives a (partial) converse to Proposition 5.14.

Proposition 5.18. *Suppose X is a Hausdorff locally convex space and suppose X is a quasi-complete Mackey space. Finally, suppose X^* is semireflexive. Then X and X^* are both reflexive.*

Proof. Since X^* is semireflexive and X is a Mackey space, $J_X : X \to J_X(X)$ is a topological isomorphism by Corollary 5.17 and Theorem 5.10(d). Hence, in view of Proposition 5.14 (X is reflexive $\Rightarrow X^*$ is reflexive), it suffices to show that J_X is onto.

Suppose $\Phi \in X^{**}$. Choose a typical neighborhood $E = B^\circ$ of zero in X^* (with B being bounded) for which $|\Phi(f)| \le 1$ when $f \in E$. That is, suppose $\Phi \in E^\circ$. Since $E = B^\circ = ((B^\circ)_\circ)^\circ$ (bipolar theorem), we can replace B with $(B^\circ)_\circ$, and thereby assume that B is closed, convex, and balanced, as well as bounded (Theorem 3.20(d)). This means that B is complete, since any Cauchy net in B is bounded, hence is convergent (since X is quasi-complete) to an element of B (since B is closed). Thus, $J_X(B)$ is complete in X^{**} since J_X is a topological isomorphism with its image, so $J_X(B)$ is strongly closed in X^{**}. Since $J_X(B)$ is also convex: $J_X(B)$ is weakly closed in X^{**}. Since we are assuming that X^* is semireflexive, the weak and weak-* topologies coincide on X^{**}, so $J_X(B)$ is weak-* closed, convex, and balanced in X^{**}. Hence $J_X(B) = (J_X(B)_\circ)^\circ$ by the bipolar theorem applied to $(X^{**}, $ weak-* topology). But practically by definition, $J_X(B)_\circ = B^\circ = E$, so $J_X(B) = E^\circ$. But $\Phi \in E^\circ$, so $\Phi \in J_X(B) \subset J_X(X)$. Since Φ was arbitrary, J_X is onto. $\qquad\qquad\qquad\qquad\qquad\qquad\qquad\qquad\qquad\qquad\qquad\qquad$ □

The next section will be concerned with a large class of spaces that are (among other things) reflexive.

5.4 Montel Spaces

Montel spaces are just about as far from being Banach spaces as you can get: A Banach space which is also a Montel space is necessarily finite-dimensional; see Exercise 8. However, Montel spaces do have a rich (and very useful) structure.

Definition 5.19. Suppose X is a Hausdorff locally convex space. X is called a **Montel space** when X is barreled; and closed, bounded sets are compact (original topology).

The next result sheds some light on the origin of the terminology.

Theorem 5.20 (Montel's Theorem, from Complex Analysis). *If U is a region in $\mathbb{C}$, then the Fréchet space $\mathscr{H}(U)$, consisting of holomorphic functions on U, is a Montel space.*

Remark. $\mathscr{H}(U)$ is Example III at the end of Sect. 3.7.

Proof. Suppose C is a closed, bounded subset of $\mathscr{H}(U)$. Since $\mathscr{H}(U)$ is metrizable, it suffices to show that any sequence $\langle f_n \rangle$ from C has a subsequence that converges to a function in C. But all seminorms are bounded on C (Corollary 3.31), so $\langle f_n \rangle$ is uniformly bounded on compact sets, hence has a convergent subsequence by the "classical" version of Montel's theorem. The limit of that subsequence then belongs to C since C is closed. $\qquad\qquad\qquad\qquad\qquad\qquad\qquad\qquad\qquad\qquad$ □

In fact, among the examples given at the end of Sect. 3.7, Examples I, III, IV, and V are Montel spaces, as are Examples I and III at the beginning of Sect. 3.8 (LF-spaces). See the exercises.

As noted at the end of the last section:

Proposition 5.21. *Montel spaces are reflexive.*

Proof. Suppose X is a Montel space, and suppose C is a weakly bounded, closed, convex subset of X. Then C is originally bounded (Corollary 3.31), and originally closed (the original topology is stronger), hence is originally compact since X is a Montel space. That means that C is weakly compact since the weak topology is weaker. Letting C float, X is semireflexive by Proposition 5.16. But now X is reflexive by Corollary 5.15.   □

To proceed, we need a lemma that is just too interesting to be called a lemma. (We will also need it in Chap. 6.)

Proposition 5.22. *Suppose X and Y are Hausdorff locally convex spaces, and suppose $\langle T_\alpha : \alpha \in D \rangle$ is a net in $\mathcal{L}_c(X, Y)$ that has the following properties:*

(a) There is a $T \in \mathcal{L}_c(X, Y)$ for which $\lim T_\alpha(x) = T(x)$ for all $x \in X$, and
(b) $\{T_\alpha : \alpha \in D\}$ is equicontinuous.

Then $T_\alpha \to T$ uniformly on compact sets.

Proof. Suppose K is compact in X and U is a barrel neighborhood of 0 in Y. Set

$$V = \left(\bigcap_{\alpha \in D} T_\alpha^{-1} \left(\frac{1}{3} U \right) \right) \bigcap T^{-1} \left(\frac{1}{3} U \right)$$

Then V is a barrel neighborhood of 0 in X since $\{T_\alpha : \alpha \in D\}$ is equicontinuous. The set $\{x + \text{int} V : x \in K\}$ is an open cover of K, so there exists $x_1, \ldots, x_n \in K$ such that

$$K \subset \bigcup_{j=1}^n (x_j + \text{int}(V)) \subset \bigcup_{j=1}^n (x_j + V).$$

Using the fact that D is directed [along with assumption (a)], choose $\alpha \in D$ such that $\beta \succ \alpha \Rightarrow T_\beta(x_j) - T(x_j) \in \frac{1}{3} U$ for $j = 1, \ldots, n$.

Suppose $x \in K$ and $\beta \succ \alpha$. Then there exists j such that $x - x_j \in V$, so that $T_\beta(x - x_j) \in \frac{1}{3} U$ and $T(x - x_j) \in \frac{1}{3} U$. Then $T(x_j - x) \in \frac{1}{3} U$ since U is balanced, and $T_\beta(x_j) - T(x_j) \in \frac{1}{3} U$ since $\beta \succ \alpha$. Hence

$$T_\beta(x) - T(x) = T_\beta(x - x_j) + T_\beta(x_j) - T(x_j) + T(x_j - x)$$

$$\in \frac{1}{3} U + \frac{1}{3} U + \frac{1}{3} U = U$$

since U is convex and $0 \in U$ (Proposition 3.3).   □

We can now prove:

Theorem 5.23. *The strong dual of a Montel space is another Montel space.*

Proof. Suppose X is a Montel space. Then X^* is barreled since X is reflexive (Propositions 5.12 and 5.21). It remains to show that closed, strongly bounded subsets of X^* are strongly compact.

Suppose C is a strongly closed, bounded subset of X^*. Then C is equicontinuous by theorem 4.16(b), so C_0 is a neighborhood of 0 in X, and so $(C_0)^\circ$ is weak-* compact by Theorem 3.26. Since C is strongly closed in $(C_0)^\circ$, it suffices to show that $(C_0)^\circ$ is strongly compact. To do this, it suffices to show that

$$((C_0)^\circ,\ \text{weak-* topology}) \to ((C_0)^\circ,\ \text{strong topology})$$

is continuous. This is done using nets (and Proposition 5.22).

Suppose $\langle f_\alpha : \alpha \in D \rangle$ is a net in $(C_0)^\circ$ that converges to $f \in (C_0)^\circ$ in the weak-* topology. Then $\{f_\alpha : \alpha \in D\}$ is contained in $(C_0)^\circ$, and so is equicontinuous since $((C_0)^\circ)_0 = C_0$ is a neighborhood of 0. The fact that $f_\alpha \to f$ in the weak-* topology now verifies all the hypotheses in Proposition 5.22, so $f_\alpha \to f$ uniformly on compact subsets of X.

Suppose B is bounded in X. Then B^- is closed and bounded, hence is compact since X is a Montel space. Hence $f_\alpha \to f$ uniformly on B^-, so $f_\alpha \to f$ uniformly on B, so there exists α such that $\beta \succ \alpha \Rightarrow |f_\beta(x) - f(x)| \leq 1$ whenever $x \in B$. That is, $\beta \succ \alpha \Rightarrow f_\beta - f \in B^\circ$.

The preceding verifies that $f_\alpha \to f$ in the strong topology, by letting B vary. Hence

$$((C_0)^\circ,\ \text{weak-* topology}) \to ((C_0)^\circ,\ \text{strong topology})$$

is continuous by Proposition 1.3(c).   □

We close this section with an interesting, and easy-to-prove result that has some unusual applications. See Atiyah-Bott [1] for such an application.

Proposition 5.24. *Suppose X is a Montel space, and B is a bounded subset of X. Then the original topology and the weak topology coincide on B.*

Proof. It suffices to show that

$$(B^-,\ \text{original topology}) \to (B^-,\ \text{weak topology})$$

is a homeomorphism. But this arrow is continuous; the space on the left is compact; and the space on the right is Hausdorff. Hence this arrow is a homeomorphism by standard results in point-set topology. (It maps compact sets to compact sets, so it maps closed sets to closed sets. This checks continuity of its inverse.)   □

5.5 Compact Convex Sets

Montel spaces provide a rich source of compact convex sets, but so do weak-*
topologies. The structure of such sets can often be exploited in unexpected ways.
The two main results of this section are the Krein–Milman theorem and a version of
the Kakutani fixed point theorem. An application of each will be outlined.

Suppose X is a Hausdorff locally convex space, and K is a compact convex
subset of X. A point $p \in K$ is called an **extreme point** if p does not lie internal to
a line segment in K. That is: If $p = tx + (1 - t)y$, with $x, y \in K$ and $0 < t < 1$,
then $p = x = y$ (so that the "line segment" between x and y reduces to a point).

Example. The extreme points of a disk constitute the boundary circle. The extreme
points of a convex polygon are its vertices.

Note that the definition of an extreme point does not involve the topology. This
matters in many applications. However, in the proof of the Krein–Milman theorem,
we will have to consider more general sets, and the topology does arise for them.

Again, suppose X is a Hausdorff locally convex space, and K is a compact
convex subset of X. A closed convex subset E of K will be called a *supporting
subset* if $p \in E$, $p = tx + (1 - t)y$ with $x, y \in K$ and $0 < t < 1$ implies that
$x, y \in E$. That is, if E contains an internal point of a line segment in K, then E
contains the endpoints (and so, being convex, contains that entire line segment). If
p is an extreme point, then $\{p\}$ is a supporting subset, but there are plenty of others:
K itself is a supporting subset, and the sides of a convex polygon are supporting
subsets.

The Krein–Milman theorem can be proven by assembling a number of lemmas,
none of which is particularly difficult. (This approach is due to Kelley.) As in the
discussion of the properties of LF-spaces, it is easiest to simply list them as facts. In
all that follows, X is a Hausdorff locally convex space over $\mathbb{R}$, and K is a nonempty
compact convex subset of X.

1. *If E is a supporting subset of K, and F is a supporting subset of E, then F is a
   supporting subset of K.*

   If $p \in F$, and $p = tx + (1 - t)y$ for $x, y \in K$ and $0 < t < 1$, then $p \in E$ (since
   $F \subset E$) so that $x, y \in E$ (since E is a supporting subset of K). Thus, $x, y \in F$
   since F is a supporting subset of E.

2. *If $\{E_\alpha : \alpha \in \mathscr{A}\}$ is any nonempty family of supporting subsets of K, then $\bigcap E_\alpha$
   is a supporting subset of K.*

   If $p \in \bigcap E_\alpha$, and $p = tx + (1 - t)y$ for $x, y \in K$ and $0 < t < 1$, then for all
   $\alpha : p \in E_\alpha$, so that $x, y \in E_\alpha$ (since E_α is a supporting set). Hence $x, y \in \bigcap E_\alpha$.
   Finally, $\bigcap E_\alpha$ is closed.

3. *If $f \in X^*$, and $M = \max f(K)$, then $\{x \in K : f(x) = M\}$ is a supporting
   subset of K.*

If $p \in K$, $f(p) = M$, and $p = tx + (1 - t)y$ with $x, y \in K$ and $0 < t < 1$, then $f(p) = tf(x) + (1 - t)f(y)$. If either $f(x) < M$ or $f(y) < M$, then $tf(x) + (1 - t)f(y) < M$ since $f(x) \leq M$ and $f(y) \leq M$. Hence $f(x) = f(y) = M$. Finally, the set is trivially closed.

4. *If K consists of more than one point, then K contains a proper nonempty supporting subset.*

Just take an $f \in X^*$ that is not constant on K, and apply Fact 3.

5. *A minimal (under set inclusion) nonempty supporting subset of K has the form $\{p\}$, for an extreme point p.*

If E is a supporting subset that consists of more than one point, then E has a proper nonempty supporting subset F by Fact 4, and this F will be a (smaller) nonempty supporting subset of K by Fact 1.

6. *K has an extreme point.*

Let

$$\mathscr{E} = \{E \subset K : E \text{ is a nonempty supporting subset of } K\},$$

and partially order $\mathscr{E}$ by (reverse) set inclusion. The set $K \in \mathscr{E}$, so $\mathscr{E}$ is nonempty. If $\{E_\alpha : \alpha \in \mathscr{A}\}$ is a nonempty chain in $\mathscr{E}$, then $E = \bigcap E_\alpha$ is a supporting set by Fact 2, and E is nonempty since K is compact. Zorn's lemma now says that $\mathscr{E}$ has a minimal element, which has the form $\{p\}$ for an extreme point p by Fact 5.

Theorem 5.25 (Krein–Milman). *Suppose X is a Hausdorff locally convex space over $\mathbb{R}$, and K is a compact convex subset of X. Then K is the closed convex hull of its set of extreme points.*

Proof. Let L denote the closed convex hull of the set of extreme points of K. Then L is closed and convex, and $L \subset K$, so L is compact. Suppose $p \in K - L$. Then there exists $f \in X^*$ and $r_0 \in \mathbb{R}$ for which $f(x) < r_0$ when $x \in L$, and $f(p) > r_0$, by Proposition 3.12. Let $M = \max f(K)$. Then $M > r_0$. Set $E = \{x \in K : f(x) = M\}$. Then E is a nonempty supporting set in K, so E has an extreme point q by Fact 6. But now $\{q\}$ is a supporting subset of E, and so is a supporting subset of K by Fact 1. Hence q is an extreme point of K, so that $q \in L$. But this is impossible, since $q \in E \Rightarrow f(q) = M > r_0 > f(x)$ whenever $x \in L$. Hence $K = L$. $\square$

Remark. In some cases, the Krein–Milman theorem can be generalized. See Phelps [28] for a discussion of this.

Application: The Stone–Weierstrass Theorem: $\mathbb{R}$ *version: Suppose X is a compact Hausdorff space, and $\mathscr{A}$ is a subalgebra of $C(X)$ that separates points; that is, if $x \neq y$ in X, then there exists $f \in \mathscr{A}$ such that $f(x) \neq f(y)$. Then $\mathscr{A}$ is either dense in $C(X)$, or $\mathscr{A}^- = \{f \in C(X) : f(p) = 0\}$ for some $p \in X$.*

Proof Outline: To start with, suppose I is an ideal in $C(X)$. Let $A = \{x \in X : f(x) = 0 \text{ for all } f \in I\}$. If B is closed and disjoint from A, then for all $x \in B$ there exists $f \in I$ s.t. $f(x) \neq 0$. Choose such an f, and set $V_x = \{y \in X : f(y) \neq 0\}$.

These V_x cover B, so since B is compact, there exists $x_1 \cdots x_n$ such that $B \subset \bigcup V_{x_j}$. Letting f_j denote the function chosen for x_j, Σf_j^2 belongs to I, is nonnegative, and is strictly positive on B. Let m be the minimum value of Σf_j^2 on B, and set $g(x) = \max(m, \Sigma f_j^2(x))$. Then g is continuous on X and is strictly positive. Hence

$$h(x) = \frac{1}{g(x)} \sum_{j=1}^{n} f_j^2(x)$$

belongs to I since $\frac{1}{g}$ belongs to $C(X)$ and I is an ideal. This $h(x)$ is easily checked to be a Urysohn function which is 0 on A and 1 on B, and has values in the interval $[0, 1]$. Finally, if $f \in C(X)$ and f vanishes on A, set $B_n = \{x \in X : |f(x)| \geq \frac{1}{n}\}$, and choose a Urysohn function $g_n \in I$ which is 0 on A and 1 on B_n, with values in $[0, 1]$. Then it is easily checked that $\|f - fg_n\| \leq \frac{1}{n}$, and $fg_n \in I$, so that $fg_n \to f$ and $f \in I^-$. Put together, this shows that $I^- = \{f \in C(X) : f|_A = 0\}$. In particular, all closed ideals in $C(X)$ have this form for some closed subset $A \subset X$.

Now suppose $\mathscr{A}$ is not dense in $C(X)$. Let K denote the intersection of $\mathscr{A}^\perp$ with the closed unit ball in X^*; K is weak-* compact and convex. Let φ be an extreme point of K. Then $\|\varphi\| = 1$ since $\mathscr{A}^\perp \neq \{0\}$.

If $f \in C(X)$, define φ_f by $\varphi_f(g) = \varphi(fg)$. $\varphi_f \in \mathscr{A}^\perp$ if $f \in \mathscr{A}$ since $\mathscr{A}$ is an algebra; also, $\varphi_c \in \mathscr{A}^\perp$ if c is constant. Note that $\varphi_{cf} = c\varphi_f$ if c is constant, and $\varphi_{f+g} = \varphi_f + \varphi_g$ since φ is linear. Suppose $g \in \mathscr{A}$. Choose $M > \|g\|$; then

$$\varphi = \varphi_1 = \varphi_{(\frac{1}{2} + \frac{g}{2M})} + \varphi_{(\frac{1}{2} - \frac{g}{2M})} \text{ and}$$

$$\|\varphi\| = \|\varphi_{(\frac{1}{2} + \frac{g}{2M})}\| + \|\varphi_{(\frac{1}{2} - \frac{g}{2M})}\| :$$

The reason for the latter is that $\frac{1}{2}\left(1 \pm \frac{g}{M}\right) \geq 0$, and

$$\|\varphi_f\| = \int_X |f| \, d|\mu|$$

when the signed measure μ represents φ. (The Riesz representation theorem is used here.) But if $\varphi = \psi + \eta$ and $\|\varphi\| = \|\psi\| + \|\eta\|$, with $\psi, \eta \in \mathscr{A}^\perp$, then either $\psi = 0$ and $\varphi = \eta$, $\eta = 0$ and $\varphi = \psi$, or $\eta \neq 0 \neq \psi$. In the latter case, set $\psi_0 = \psi/\|\varphi\|$, $\eta_0 = \eta/\|\eta\|$, and $t = \|\varphi\|$. Then $\varphi = t\psi_0 + (1 - t)\eta_0$ with $\psi_0, \eta_0 \in K$, so that $\psi_0 = \eta_0 = \varphi$ since φ is an extreme point. In all cases, ψ and η are scalar multiples of φ. Unraveling φ-sub-$\left(\frac{1}{2} \pm \frac{1}{2}g/M\right)$, $\varphi_g = c(g)\varphi$ for some scalar $c(g)$ when $g \in \mathscr{A}$.

Now let $I = \{g \in C(X) : \varphi_g = 0\}$. Then I is a closed proper ideal in $C(X)$ (easy check), so $I = \{f \in C(X) : f|_A \equiv 0\}$ for some closed set $A \subset X$. However, if $f \in \mathscr{A}$, then $f - c(f) \in I$, so $f(a) = c(f)$ for all $a \in A$. Since $\mathscr{A}$ separates points, $A = \{p\}$ consists of one point, so $I = \{f \in C(X) : f(p) = 0\}$. Since $C(X)/I$ is one-dimensional and φ vanishes on I, φ is $\pm$ evaluation at p. Since φ vanishes on $\mathscr{A}$, every function in $\mathscr{A}$ vanishes at p. Since $\mathscr{A}$ separates points, this p is unique, so $\pm\varphi$ are the only extreme points of K. By the Krein–Milman theorem, $K = [-1, 1]\varphi$, $\mathscr{A}^\perp = \mathbb{R}\varphi$, and $\mathscr{A}^- = \{f \in C(X) : f(p) = 0\}$. □

This approach is due to de Branges [8]. $\mathbb{C}$-*version: Suppose* X *is a compact Hausdorff space, and* $\mathscr{A}$ *is a* $*$-*subalgebra of* $C(X)$, *here denoting continuous complex-valued functions on* X, *while* $*$-*subalgebra denotes a subalgebra* $\mathscr{A}$ *for which* $f \in \mathscr{A} \Rightarrow f^* \in \mathscr{A}$, *where* $f^*(x) = \overline{f(x)}$. *If* $\mathscr{A}$ *separates points, then* $\mathscr{A}$ *is either dense in* $C(X)$, *or* $\mathscr{A}^- = \{f \in C(X) : f(p) = 0\}$ *for some* $p \in X$.

Proof Outline: Let $\mathscr{A}_0 = \{f \in \mathscr{A} : f$ is real-valued$\}$, and $C_0(X) = \{f \in C(X) : f$ is real valued$\}$. If $f \in \mathscr{A}$, then Re$f = \frac{1}{2}(f + f^*)$ and Im$f = \frac{i}{2}(f^* - f)$ belong to $\mathscr{A}_0$, so $\mathscr{A}_0$ is a subalgebra of $C_0(X)$ that separates points. Hence by the $\mathbb{R}$-version, $\mathscr{A}_0^-$ is either $C_0(X)$ (in which case $\mathscr{A}^- = C(X)$) or $\mathscr{A}_0^- = \{f \in C_0(X) : f(p) = 0\}$ for some $p \in X$ (in which case $\mathscr{A}^- = \{f \in C(X) : f(p) = 0)$. □

Example 1. $X =$ closed unit disk in $\mathbb{C}$, and

$$\mathscr{A} = \{f \in C(X) : f \text{ is holomorphic on int}(X)\}.$$

$\mathscr{A}$ is a closed subalgebra in $C(X)$ that separates points, is not dense, and has no common zero; but $\mathscr{A}$ is not a $*$-subalgebra,

The following result is useful in several applications of the Krein–Milman theorem, including the Kakutani fixed-point theorem.

Proposition 5.26 (Milman). *Suppose* X *is a Hausdorff locally convex space, and* K *is a compact subset of* X *with a closed convex hull* con$(K)^-$ *that is also compact. Then all extreme points of* con$(K)^-$ *belong to* K.

Proof. Suppose not; suppose p is an extreme point of con$(K)^-$ that is not in K. Let U be a barrel neighborhood of 0 for which $(p + U) \cap K = \emptyset$. The set of all $x + $ int(U), with $x \in K$, covers K. Hence there exists $x_1 \cdots x_n \in K$ for which $K \subset \bigcup(x_j + \text{int}U)$ since K is compact. Set $A_j = \text{con}(K \cap (x_j + U))^-$, a closed convex subset of con$(K)^-$. Thus each A_j is compact. If C and D are compact and convex, then con$(C \bigcup D)$ is compact by Proposition 2.14 (with $I = [0, 1]$), since con$(C \bigcup D)$ is a continuous image of $C \times D \times I$, a compact set. Hence by induction on n, con$(A_1 \cup \cdots \cup A_n)$ is compact. It also contains K, so con$(A_1 \cup \cdots \cup A_n) \supset$ con$(K)^-$. Thus, $p \in$ con$(A_1 \cup \cdots \cup A_n)$. Write $p = \sum t_j a_j$ for $a_j \in A_j$ and $0 \le t_j \le 1, \sum t_j = 1$. If $0 < t_j < 1$, then

$$p = t_j a_j + (1 - t_j) \sum_{i \ne j} \frac{t_i}{1 - t_j} a_i$$

so that $p = a_j$ since p is extreme and all $a_j \in A_j \subset \text{con}(K)^-$. If $t_j = 1$, then all other $t_i = 0$ and $p = a_j$. In all cases, $p = a_j \in A_j$ for some j, since $\sum t_j = 1 \Rightarrow$ some $t_j > 0$. But $x_j + U$ is closed and convex, so $A_j = \text{con}(K \cap (x_j + U))^- \subset x_j + U$. This means that $p \in x_j + U$, giving $p - x_j \in U$, so that $x_j - p \in U$ (U is balanced) and $x_j \in p + U$. But this means that $x_j \in (p + U) \cap K = \emptyset$. $\qquad\square$

Corollary 5.27 (Kakutani). *Suppose X is a Hausdorff locally convex space, and suppose K is a nonempty compact convex subset of X. Suppose $\mathscr{G}$ is a group of invertible members of $\mathscr{L}_c(X, X)$ with the property that for all $x \in X$, $\mathscr{G} \cdot x = \{T(x) : T \in \mathscr{G}\}$ is compact. Finally, suppose that $T(K) \subset K$ for all $T \in \mathscr{G}$. Then there exists $x \in K$ such that $T(x) = x$ for all $T \in \mathscr{G}$.*

Remark. Such an x is called a *fixed point* of $\mathscr{G}$.

Proof. Let

$$\mathscr{D} = \{E \subset K : E \neq \emptyset, E \subset K, E \text{ is}$$

$$\text{closed and convex, and } T(E) \subset E \text{ for}$$

$$\text{all } T \in \mathscr{G}\}.$$

$K \in \mathscr{D}$, so $\mathscr{D}$ is nonempty. If $\mathscr{C}$ is a nonempty chain in $\mathscr{D}$, then $\bigcap \mathscr{C} \in \mathscr{D}$: The intersection is nonempty since K is compact, while the intersection is closed and convex and $\mathscr{G}$-invariant: If $x \in \bigcap \mathscr{C}$, then for all $E \in \mathscr{C} : x \in E$. If $T \in \mathscr{G}$, then $T(x) \in E$. This holds for all $E \in \mathscr{C}$, so $T(x) \in \bigcap \mathscr{C}$. The preceding shows that $\mathscr{D}$ has a minimal (under set inclusion) element E by Zorn's lemma.

Suppose $x \in E$. Then $\mathscr{G} \cdot x = \{T(x) : T \in \mathscr{G}\}$ is compact and contained in E, so $\text{con}(\mathscr{G} \cdot x)^- \subset E$ since E is closed and convex. Since $\mathscr{G}$ is closed under composition and consists of linear transformations, $\mathscr{G}$ preserves $\text{con}(\mathscr{G} \cdot x)$:

$$T(\Sigma t_j T_j(x)) = \Sigma t_j (T T_j)(x).$$

Finally, since $\mathscr{G}$ is group, each $T \in \mathscr{G}$ is a homeomorphism, so $T(A^-) = T(A)^-$ for any set A. In particular, $T(\text{con}(\mathscr{G} \cdot x)^-) = T(\text{con}(\mathscr{G} \cdot x))^- \subset \text{con}(\mathscr{G} \cdot x)^-$. What all this shows is that $\text{con}(\mathscr{G} \cdot x)^- \in \mathscr{D}$. But we also know that $\mathscr{G} \cdot x \subset E$, so that $\text{con}(\mathscr{G} \cdot x) \subset E$ (since E is convex) and $\text{con}(\mathscr{G} \cdot x)^- \subset E$ (since E is closed). Since E is minimal, $E = \text{con}(\mathscr{G} \cdot x)^-$.

Suppose $x, y \in E$. Set $z = \frac{1}{2}x + \frac{1}{2}y$. Then $E = \text{con}(\mathscr{G} \cdot z)^-$. E has an extreme point (Krein–Milman) that belongs to $\mathscr{G} \cdot z$ since $\mathscr{G} \cdot z$ is compact (Proposition 5.26). But $T(z) = \frac{1}{2}T(x) + \frac{1}{2}T(y)$, so $T(z) = T(x) = T(y)$ when $T(z)$ is extreme. Applying T^{-1}, $x = z = y$. Since x and y were arbitrary in E, E must consist of one point. Since $T(E) \subset E$, that point is a fixed point. $\qquad\square$

Remark. A more general version is Theorem 5.11 of Rudin [32].

Application. Haar Measures for Compact Hausdorff Topological Groups.
Suppose G is a compact Hausdorff topological group. Then there exists a Baire measure μ on X for which $\mu(G) = 1$, and $\mu(xE) = \mu(E)$ for all Baire sets E and $x \in G$.

Proof Outline: Let $X = C(G)$, real-valued continuous functions on G. The locally convex space to work with is X^*, which corresponds (under Riesz representation) to signed finite Baire measures on G. The set K is the subset of X^* corresponding to positive measures μ for which $\mu(G) = 1$. This is easily checked to be a closed convex subset of the closed unit ball in X^*, and so is weak-* compact. To define $\mathscr{G}$, if $x \in G$ define $\pi(x) \in \mathscr{L}_c(X, X)$ as follows:

$$(\pi(x)f)(y) = f(x^{-1}y). \qquad\qquad \text{Then}$$

$$(\pi(x)\pi(y)f)(z) = [\pi(x)(\pi(y)f)](z)$$
$$= (\pi(y)f)(x^{-1}z) = f(y^{-1}x^{-1}z)$$
$$= f((xy)^{-1}z) = (\pi(xy)f)(z),$$

so $\pi(xy) = \pi(x)\pi(y)$. Each $\pi(x)$ is easily checked to be an isometry onto with inverse $\pi(x^{-1})$, so each $\pi(x)$ is continuous. $\mathscr{G} = \{\pi(x)^* : x \in G\}$ is now a group, and a $\mathscr{G}$-fixed point in K is easily checked to be a Haar measure. (Each $\pi(x)^*$ is continuous when X^* has the weak-* topology thanks to Theorem 5.2.) $\pi(x)^*K \subset K$ when $x \in G$, so the existence of a Haar measure will follow once it is shown that for all $\varphi \in X^*$, $\mathscr{G} \cdot \varphi$ is compact. This follows from the fact that $g \mapsto \pi(g)^*\varphi$ is continuous from G to $(X^*, \text{weak-* topology})$.

To see this, for the usual reasons involving subbases, it suffices to show that if $f \in C(X)$, then there exists a neighborhood V of 1 in G for which

$$|(\pi(x)^*\varphi - \varphi)(f)| \leq 1 \text{ when } x \in V, \text{ so that } \pi(x)^*\varphi - \varphi \in \{f\}^\circ.$$

But $(\pi(x)^*\varphi)(f) = \varphi(\pi(x)f)$, so this reads "$|\varphi(\pi(x)f - f)| \leq 1$." To do this, it suffices to show (since φ is bounded) that given $\varepsilon > 0$, there is a neighborhood V of 1 in G such that $\|\pi(x)f - f\| < \varepsilon$ when $x \in V$. This is really uniform continuity for f.

Define $g : G \times G \to \mathbb{R}$ by

$$g(x, y) = f(x^{-1}y) - f(y).$$

g is continuous, and $g(1, y) = 0$ for all $y \in G$. For all $y \in G$, there exist open neighborhoods V_y of 1 and W_y of y for which

$$|g(x, z)| < \varepsilon \text{ when } (x, z) \in V_y \times W_y$$

since g is jointly continuous at $(1, y)$. The sets W_y cover G, so there exists $y_1, \ldots, y_n$ for which $G \subset \bigcup W_{y_k}$. Set $V = \bigcap V_{y_k}$. If $y \in G$ and $x \in V$, then there exists k for which $y \in W_k$. But $x \in V_k$, so $|g(x, y| < \varepsilon$. Hence $|g(x, y)| < \varepsilon$ whenever $x \in V$ and $y \in G$. $\qquad\qquad\qquad\qquad\qquad\qquad \square$

Haar measures here are defined to be left invariant. For compact groups, they are also right invariant, but this does not hold for more general locally compact groups. For a discussion of all this (including a version of uniqueness of Haar measures), consult any good book about locally compact groups.

5.6   Ptak's Closed Graph Theorem

A closed graph theorem has already appeared in Sect. 4.5, with the hypotheses that the domain space is barreled and the range space is a Fréchet space. The assumption about the range space is particularly restrictive; there should be (and there is) some way to generalize it. The subject of this section is to see how this can be done using dual spaces. Before proceeding, however, it seems worthwhile to verify that the hypothesis imposed on the domain space (i.e., "barreled") really is the right one. The next result does that.

Proposition 5.28 (Mahowald). *Suppose* X *is a Hausdorff locally convex space with the following property:*

> *Whenever* Y *is a Banach space, and* $T : X \to Y$ *is a linear map with closed graph, then* T *is necessarily continuous.*

Then X *is barreled.*

Proof. Let B be a barrel in X, and let p_B denote its Minkowski functional. Let M denote the kernel of p_B:

$$M = \{x \in X : p_B(x) = 0\} = \bigcap_{n=1}^{\infty} \frac{1}{n} B.$$

Note that M is a *closed* subspace, since the intersection on the right is closed. Also, p_B is a seminorm, which induces a norm on X/M. Let Y denote a completion of X/M.

Note: One can embed any normed space Z in a completion simply by using the closure of $J_Z(Z)$ in Z^{**}. For clarity, it helps to pretend that X/M above is literally a subspace of Y. See the end of Appendix A for a mechanism that can be used to carry this out.

This Y is a Banach space. Let $T : X \to Y$ be defined in the obvious way: $T(x) = x + M$, so that the graph $\Gamma(T)$ of T is

$$\Gamma(T) = \{(x, x + M) \in X \times Y : x \in X\}.$$

This T is linear. It also has closed graph:

Suppose $\langle (x_\alpha, x_\alpha + M) : \alpha \in D \rangle$ is a net in $\Gamma(T)$ that converges to $(x, y) \in X \times Y$. Then $x_\alpha \to x$ in X, and $\langle x_\alpha + M \rangle$ is a Cauchy net relative to the norm p_B; that is, $\langle x_\alpha \rangle$ is a Cauchy net relative to the seminorm p_B. For each n, choose $\alpha_n \in D$ so that $\beta, \gamma \succ \alpha_n \Rightarrow p_B(x_\beta - x_\gamma) < \frac{1}{n}$. Then $\beta, \gamma \succ \alpha_n \Rightarrow x_\beta - x_\gamma \in \frac{1}{n} B$ by Theorem 3.7.

Suppose $\beta \succ \alpha_n$. If U is any convex, balanced neighborhood of 0 in X, choose $\delta \in D$ such that $\gamma \succ \delta \Rightarrow x_\gamma - x \in U$. Next, choose $\gamma \succ \delta$ and $\gamma \succ \alpha_n$, using the fact that D is directed. Then

$$x_\beta - x = (x_\beta - x_\gamma) + (x_\gamma - x) \in \frac{1}{n} B + U.$$

Letting U vary, this shows that $x_\beta - x \in \frac{1}{n} B + U$ for all convex balanced neighborhoods U of 0, so $x_\beta - x \in \frac{1}{n} B$ since $\frac{1}{n} B$ is closed (Proposition 1.9). In particular, $p_B(x_\beta - x) \le \frac{1}{n}$ when $\beta \succ \alpha_n$. That simply means that $x_\beta + M \to x + M$ in $(X/M, p_B)$, that is that $y = x + M$ by uniqueness of limits in Y.

So: T has closed graph, so T must be continuous. But

$$T^{-1}(\{x + M \in X/M : p_B(x) < 1\}) \subset B$$

by Theorem 3.7, so B is a neighborhood of 0 in X. $\square$

Now to what can be said using dual spaces. Suppose X and Y are Hausdorff locally convex spaces, and $T : X \to Y$ is *any* linear transformation. Define the adjoint T^*, with domain $D(T^*) \subset Y^*$, as follows:

$$D(T^*) = \{f \in Y^* : f \circ T \in X^*\},$$
$$T^*(f)(x) = f(T(x))$$

It is easily checked that $D(T^*)$ is a subspace of Y^*, and (as in Sect. 5.1) $T^* : D(T^*) \to X^*$ is a linear transformation. Somewhat more can be said when T has closed graph.

Proposition 5.29. *Suppose X and Y are Hausdorff locally convex spaces, and $T : X \to Y$ is a linear transformation. Letting $\Gamma(?)$ denote the graph, and identifying $(X \times Y)^*$ with $Y^* \times X^*$ via Theorem 5.9 (note the reversal of order):*

$$\Gamma(-T^*) = \Gamma(T)^\perp.$$

Finally, if $\Gamma(T)$ is closed, then $D(T^*)$ is weak-* dense in Y^*.

Proof. Observe that $(f, g) \in \Gamma(T)^\perp$ if, and only if, $f(x) + g(T(x)) = 0$ for all $x \in X$. That is, if and only if $g(T(x)) = -f(x)$. Since f is continuous, automatically $g \in D(T^*)$ and $T^*(g) = -f$, that is $(-T^*)(g) = f$. On the other hand, if $g \in D(T^*)$ and $f = -T^*(g)$, then $f(x) + g(T(x)) = f(x) + T^*(g)(x) = 0$ for all $x \in X$. When put together, this all shows that $\Gamma(-T^*) = \Gamma(T)^\perp$.

Now suppose $\Gamma(T)$ is closed. Note that $\Gamma(T)$ is a subspace of $X \times Y$ since T is linear. If $y \in Y$ and $y \ne 0$, then $(0, y) \notin \Gamma(T)$, so there exists a continuous linear functional on $(X \times Y)/\Gamma(T)$ that does not vanish at $(0, y) + \Gamma(T)$. By Theorem 5.7, this linear functional is represented by some $(f, g) \in \Gamma(T)^\perp$, that is it has the form $(\pi^*)_0^{-1}(-T^*(g), g)$ for $g \in D(T^*)$ (in the notation of Theorem 5.7), thanks to the fact that $\Gamma(T)^\perp = \Gamma(-T^*)$. In particular, $y \notin D(T^*)_\perp$ since $g(y) \ne 0$ for this g. Letting y vary, $D(T^*)_\perp = \{0\}$.

Let Z denote the weak-* closure of $D(T^*)$. Then $Z \neq Y^* \Rightarrow$ there exists a weak-* continuous, nonzero linear functional on Y^* that vanishes on Z (Hahn–Banach theorem). But any such functional is evaluation at a point $y \in Y$ by Corollary 3.23(b), and that is excluded by the preceding. Hence $Z = Y^*$, and $D(T^*)$ is weak-* dense in Y^*. □

Now it becomes clear what *kind* of assumption we should make on the range space. The domain space will be assumed to be barreled, hence infrabarreled, so Proposition 4.39 will guarantee that a linear map with a closed graph will be continuous provided we make some kind of assumption on the range space that will force the weak-* dense $D(T^*)$ to be all of Y^*. Ptak came up with just such an assumption.

Definition 5.30. Suppose Y is a Hausdorff locally convex space. Y is **B-complete** if any subspace M of Y^* with the property that:

$$M \bigcap U^\circ \text{ is weak-* closed in } Y^*$$

for all barrel neighborhoods U

of 0 in Y,

is weak-* closed in Y^*. Y is **B_r-complete** if the preceding holds whenever M is weak-* dense in Y^* (in which case the conclusion is that $M = Y^*$).

Clearly, B-complete spaces are B_r-complete. Also, Fréchet spaces are B-complete; this is normally proven using Krein–Smulian I in Sect. 6.2. However, Ptak proved that for barreled spaces, B_r-completeness is the "right" assumption:

If X is a barreled Hausdorff locally convex space with the property that any linear map from a barreled space to X with closed graph is continuous, then X is B_r-complete.

This is Exercise 20; it has a number of steps, none of which is particularly difficult. Since Fréchet spaces *are* appropriate targets for a closed graph theorem (and are barreled), it follows that Fréchet spaces are B_r-complete. In fact, looking at quotient spaces, Fréchet spaces are B-complete (Exercise 21). This approach to proving that Fréchet spaces are B-complete does not seem to be well known. It should be.

By the way, Exercise 31 of Chap. 4 establishes a closed graph theorem for maps from one LF-space to another. LF-spaces need not be B_r-complete, but the domain is more restricted there.

Before proceeding to Ptak's closed graph theorem (and its companion open mapping theorem), a few results about B-completeness and B_r-completeness are in order. First of all, the word "complete" following B and B_r really is justified.

Proposition 5.31. B_r-*complete spaces are complete.*

Proof. Suppose X is B_r-complete, and suppose $\langle x_\alpha : \alpha \in D \rangle$ is a Cauchy net in X. If $f \in X^*$, then for all $\varepsilon > 0$ there exists a barrel neighborhood U of 0 in X for which $x \in U \Rightarrow |f(x)| < \varepsilon$. But now there exists $\alpha \in D$ for which $\beta, \gamma \succ \alpha \Rightarrow x_\beta - x_\gamma \in U$, so that $|f(x_\beta) - f(x_\gamma)| = |f(x_\beta - x_\gamma)| < \varepsilon$. That is, $\langle f(x_\alpha) : \alpha \in D \rangle$ is a Cauchy net in the base field, hence must be convergent there. Set

$$\Phi(f) = \lim_\alpha f(x_\alpha) \text{ for } f \in X^*.$$

It is easily checked that Φ is a linear functional on X^*, so its kernel M has codimension 1 in Y^*. Hence M is either closed in Y^* or is dense in Y^*.

Suppose M is dense in Y^*, and U is a barrel neighborhood of 0. For each $\varepsilon > 0$, there exists $\alpha \in D$ such that $\beta, \gamma \succ \alpha \Rightarrow x_\beta - x_\gamma \in \varepsilon U$. That is, $\varepsilon^{-1}(x_\beta - x_\gamma) \in U$. If $f \in U^\circ$, then $\varepsilon^{-1}|f(x_\beta) - f(x_\gamma)| = |f(\varepsilon^{-1}(x_\beta - x_\gamma))| \le 1$, so $|f(x_\beta) - f(x_\gamma)| \le \varepsilon$. Since $\varepsilon > 0$ is arbitrary, this all shows that the net limit

$$\Phi(f) = \lim_\alpha f(x_\alpha)$$

is uniform on U°. Since each evaluation $f \mapsto f(x_\alpha)$ is weak-* continuous, the uniform limit $f \mapsto \Phi(f)$ is also weak-* continuous. (See Appendix A if this net-limit property of limits is unfamiliar.) That is, Φ is weak-* continuous on U°, so $M \bigcap U^\circ = \{f \in U^\circ : \Phi(f) = 0\}$ is closed. Since X is B_r-complete and U is arbitrary, M is closed. That is, M *is* weak-* *closed whether it is* weak-* *dense or not*, so Φ is weak-* continuous. Hence Φ is given by evaluation at a point $x \in X$ by Corollary 3.23(b). It remains to show that $x_\alpha \to x$ in X.

Suppose U is a barrel neighborhood of 0 in X. There exists $\alpha \in D$ such that $\beta, \gamma \succ \alpha \Rightarrow x_\beta - x_\gamma \in U$, so that for all $f \in U^\circ$: $|f(x_\beta) - f(x_\gamma)| = |f(x_\beta - x_\gamma)| \le 1$. Freeze β, and let $\gamma \to \infty$: $\{\gamma \in D : \gamma \succ \alpha\}$ is cofinal in D since D is directed, so the limit of $f(x_\gamma)$ over this set is also $\Phi(f) = f(x)$ by Proposition 1.4. Thus $|f(x_\beta - x)| = |f(x_\beta) - f(x)| \le 1$ as well since U is closed. Letting f vary over U° (but keeping x_β fixed), $x_\beta - x \in (U^\circ)_\circ$. But $U = (U^\circ)_\circ$ by the bipolar theorem, so (now letting β vary) $\beta \succ \alpha \Rightarrow x_\beta - x \in U$. This is convergence of $\langle x_\alpha \rangle$ to x when U is also allowed to vary. $\square$

There is one more result before getting to the closed graph and open mapping theorems.

Proposition 5.32. *Suppose X is a Hausdorff locally convex space, and Y is a closed subspace. If X is B-complete, then so is X/Y.*

Proof. This starts with the identification $(X/Y)^* \approx Y^\perp$. The first point is that this algebraic isomorphism is actually a topological isomorphism when weak-* topologies are used. Letting $\pi : X \to X/Y$ denote the natural map, note that if F is finite in X, then $\pi(F)$ is finite in X/Y, and all finite subsets of X/Y arise in this way. Furthermore, if F is finite in X, then

$$F^\circ \bigcap Y^\perp = \{f \in Y^\perp : |f(x)| \le 1 \text{ for all } x \in F\},$$

while the polar of $\pi(F)$ in Y^* is

$$\{f \in Y^\perp : |[(\pi^*)_0^{-1}(f)](x + Y)| \leq 1\}.$$

But $[(\pi^*)_0^{-1}(f)](x + Y) = f(x)$: Set $g = (\pi^*)_0^{-1}(f)$, so that $(\pi^*)_0(g) = f$, that is $\pi^*(g) = f$. Then $f(x) = [\pi^*(g)](x) = g(\pi(x)) = g(x + Y)$ as stated above. (This is messy to make explicit, but is still direct.)

What all this shows is that the weak-* topology on X induces the (X/Y) – weak-* topology on $Y^\perp$.

Suppose M is a subspace of $Y^\perp$ for which $M \cap U^\circ$ is weak-* closed whenever U is a barrel neighborhood of 0 in X/Y. If V is a barrel neighborhood of 0 in X, then $\pi(\text{int}(V))$ is open in X/Y since π is an open mapping, so there exists a barrel neighborhood U of 0 in X/Y such that $U \subset \pi(\text{int}(V)) \subset \pi(V)$. Using the fact that for $f \in Y^\perp$: $f(x)$ is identified as $f(x + Y) = f(\pi(x))$, $|f(\pi(x))| \leq 1$ exactly when $|f(x)| \leq 1$ (see above), so

$$\underset{\substack{\uparrow \\ \text{polar} \\ \text{in } X^*}}{V^\circ} \cap Y^\perp = \underset{\substack{\uparrow \\ \text{polar in} \\ Y^\perp \approx (X/Y)^*}}{\pi(V)^\circ} \subset U^\circ.$$

Hence

$$M \cap V^\circ = (M \cap Y^\perp) \cap V^\circ \qquad (\text{since } M \subset Y^\perp)$$

$$= M \cap \pi(V)^\circ = (M \cap U^\circ) \cap \pi(V)^\circ,$$

an intersection of two weak-* closed subsets of $Y^\perp$. Since X is B-complete and V is arbitrary, M is weak-* closed in X^*, and so is weak-* closed in $Y^\perp$.

$\square$

Now for Ptak's closed graph and open mapping theorems. Ptak's results are more general, but at the expense of a much more complicated description.

Theorem 5.33 (Closed Graph Theorem of Ptak). *Suppose X and Y are Hausdorff locally convex spaces, and suppose X is barreled and Y is B_r-complete. If $T : X \to Y$ is a linear map whose graph, $\Gamma(T)$, is closed in $X \times Y$, then T is continuous.*

Proof. In view of Proposition 4.39, since X is infrabarreled, it suffices to show that $D(T^*) = Y^*$. Since $D(T^*)$ is weak-* dense in Y^* (Proposition 5.29), it suffices (since Y is assumed to be B_r-complete) to show that $D(T^*) \cap U^\circ$ is weak-* closed whenever U is a barrel neighborhood of 0 in Y. This is done by showing that $D(T^*) \cap U^\circ$ is the continuous image of a compact set.

Let $\pi : X^* \times Y^* \to Y^*$ denote the canonical projection, and suppose U is a barrel neighborhood of 0 in Y. Then $T^{-1}(U)$ is convex, balanced, and absorbent, so $T^{-1}(U)^-$ is a barrel in X, and so is a neighborhood of 0 since X is barreled. Set

$$K = \Gamma(-T^*) \bigcap [(T^{-1}(U)^-)^\circ \times U^\circ]$$

$$= \Gamma(T)^\perp \bigcap [(T^{-1}(U)^-)^\circ \times U^\circ].$$

Then K is weak-* compact in $(X \times Y)^* \approx X^* \times Y^*$ since $\Gamma(T)^\perp$ is closed and $(T^{-1}(U)^-)^\circ \times U^\circ$ is weak-* compact (Banach-Alaoglu). (*Note:* The product of the weak-* topologies on $X^* \times Y^*$ is the weak-* topology of $(X \times Y)^*$ since the ι_k^* and π_k^* mappings in Theorem 5.9 are weak-* continuous by Theorem 5.2.)

Suppose $(f, g) \in K$. Then $g \in D(T^*)$, and $T^*(g) = -f \in (T^{-1}(U)^-)^\circ$ as well as $g \in U^\circ$. In particular, $g = \pi(f, g) \in D(T^*) \bigcap U^\circ$. On the other hand, if $g \in D(T^*) \bigcap U^\circ$, and $f = -T^*(g)$, so that $(f, g) \in \Gamma(-T^*)$, then for $x \in T^{-1}(U)$:

$$|f(x)| = |-T^*(g)(x)| = |g(T(x))| \le 1$$

since $T(x) \in U$. Hence $T^{-1}(U) \subset \{x \in X : |f(x)| \le 1\}$, a closed set, so $T^{-1}(U)^- \subset \{x \in X : |f(x)| \le 1\}$. That is, $f \in (T^{-1}(U)^-)^\circ$, so $-f \in (T^{-1}(U)^-)^\circ$ since polars are balanced. In particular, $(f, g) \in K$, so $g = \pi((f, g)) \in \pi(K)$. What all this shows is that $D(T^*) \bigcap U^\circ = \pi(K)$, a weak-* compact set.   □

Corollary 5.34 (Open Mapping Theorem of Ptak). *Suppose X and Y are Hausdorff locally convex spaces, and suppose Y is barreled and X is B-complete. If $T : X \to Y$ is continuous and onto, then T is an open map.*

Proof. Replace X with $X/\ker(T)$, a space that is also B-complete, hence is B_r-complete, by Proposition 5.32. We have a continuous algebraic isomorphism

$$X/\ker(T) \xrightarrow{T_0} Y$$

by Theorem 1.23(c); to show that T is an open map, it suffices to show that $(T_0)^{-1}$ is continuous. But T_0 is continuous, so $\Gamma(T_0)$ is closed. Hence $\Gamma(T_0^{-1})$ is closed by the exchange $(X/\ker(T)) \times Y \approx Y \times (X/\ker(T))$. Hence T_0^{-1} is continuous by Ptak's closed graph theorem.   □

See Exercise 19 for another class of B-complete spaces.

If it were not for the results of this section, most of the next section would belong in the next chapter. However, the results here broaden things quite a bit.

5.7   Closed Range Theorems

The fact that the image of a transformation behaves more poorly than the kernel is familiar from group theory: The kernel of a homomorphism is a normal subgroup, while its range is a subgroup that is not necessarily normal. Something similar happens in ring theory: Kernels are ideals, while images are only subrings. In functional analysis, what fails is not algebraic, but topological: The kernel of a continuous linear transformation from one Hausdorff locally convex space to another is a closed subspace, but its range is a subspace that need not be closed.

By the way, the words "image" and "range" were used as synonyms in the preceding paragraph, but the word "range" is the one normally used in the context of "closed range":

Definition 5.35. Suppose X and Y are Hausdorff locally convex spaces, and $T : X \to Y$ is a continuous linear transformation. Then T has **closed range** when $T(X)$ is a closed subspace of Y.

Some of what matters here appears in disguise in Theorem 5.2. The following is simply a clarification.

Theorem 5.36. *Suppose X and Y are Hausdorff locally convex spaces, and $T : X \to Y$ is a continuous linear transformation. Then*

(a) $T(X)^{\perp} = \ker T^*$,
(b) $(\ker T^*)_{\perp} = T(X)^-$,
(c) $T^*(Y^*)_{\perp} = \ker T$, and
(d) $(\ker T)^{\perp}$ is the weak-* closure of $T^*(Y^*)$.

Proof. In the third sentence of Theorem 5.2, setting $A = X$ gives $T(X)^\circ = (T^*)^{-1}(X^\circ) = \ker T^*$; but $T(X)^\circ = T(X)^{\perp}$ since $T(X)$ is a subspace. This gives (a); it also gives (c) by using weak-* topologies on the dual spaces.

As for (b): $(T(X)^{\perp})_{\perp} = (\ker T^*)_{\perp}$ by part (a). But $(E^{\perp})_{\perp} = E^-$ for subspaces by the bipolar theorem, so $(\ker T^*)_{\perp} = T(X)^-$. Part (d) now follows by again using weak-* topologies. $\qquad\square$

Moral: If all maps in Theorem 5.36 have closed range, then there is a nice symmetry between (a) and (b), and between (c) and (d). Nice symmetries are not enough, however. The utility of "closed range" goes far beyond that. One example is the formation of homology spaces: If $d : X \to Y$ and $\delta : Y \to Z$ are continuous linear transformations of Hausdorff locally convex spaces with $\delta d = 0$, then d has closed range exactly when the homology space $\ker \delta / d(X)$ is a Hausdorff locally convex space. This kind of thing matters. Other usages apply to solving equations: By part (b), T has closed range exactly when

$$\text{``}T(x) = y \text{ has a solution for a}$$
$$\text{fixed } y \Leftrightarrow f(y) = 0 \text{ for all } f \in \ker(T^*)\text{''}$$

is valid.

The first few theorems relate T with T^*. For the first one, the condition on the range space is a bit peculiar, but does arise in practice.

Proposition 5.37. *Suppose X and Y are Hausdorff locally convex spaces, and suppose X is B-complete (or a Fréchet space) and Y has the property that every closed subspace is barreled. If $T : X \to Y$ is a continuous linear transformation with closed range, then T^* has weak-* closed range.*

Remark. The property hypothesized for Y holds for Fréchet spaces.

Proof. Assuming $T(X)$ is closed forces $T(X)$ to be barreled, so we now have that T is an open map (Corollary 5.34 or Theorem 4.35) onto $T(X)$. Hence the induced map $T_0 : X/\ker(T) \to T(X)$ is a topological isomorphism (Theorem 1.23). Hence the induced map $T_0^* : T(X)^* \to (X/\ker(T))^* \approx \ker(T)^\perp$ is bijective. However, using the composite:

$$X \xrightarrow{\pi} X/\ker(T) \xrightarrow{T_0} T(X) \qquad\qquad T = T_0\pi$$

we get that $T^* = \pi^* T_0^* \iota^*$, where $\iota : T(X) \to Y$ is the inclusion:

$$Y^* \xrightarrow{\iota^*} T(X)^* \xrightarrow{T_0^*} (X/\ker(T))^* \xrightarrow[\approx]{\pi^*} \ker T^\perp \subset X^*$$

$$\uparrow \qquad\qquad\qquad\qquad\qquad \uparrow$$

$$\text{onto, Theorem 5.6} \qquad\qquad\qquad \text{Theorem 5.7}$$

$\square$

This has a companion.

Proposition 5.38. *Suppose X and Y are Hausdorff locally convex spaces, and suppose X is barreled and B-complete (or is a Fréchet space) and Y is first countable. If $T : X \to Y$ is a continuous linear transformation, and if T^* has weak-* closed range, then T has closed range.*

Proof. (Very weird): Assuming that T^* has weak-* closed range, we have that $T^*(Y^*) = (\ker T)^\perp$ (Theorem 5.36(d)), so that $T^*(Y^*) \approx (X/\ker(T))^*$ by Theorem 5.7. Let τ_1 be the quotient topology of $X/\ker(T)$ transported over to $T(X)$, and let τ_2 be the induced topology on $T(X)$ as a subspace of Y. The preceding shows that the dual space of $(T(X), \tau_1)$ is precisely given by $Y^*/\ker(T^*)$ (since that is isomorphic to $T^*(Y^*) = (X/\ker(T))^*$ as a vector space), while the dual of $(T(X), \tau_2)$ is given by $Y^*/T(X)^\perp$ by Theorem 5.6. But $T(X)^\perp = \ker(T^*)$ by Theorem 5.36(a), so $(T(X), \tau_1)$ and $(T(X), \tau_2)$ have the same continuous linear functionals.

Now for the weird part. $(T(X), \tau_1)$ is barreled since $X/\ker(T)$ is barreled (Proposition 4.2(a)), so $(T(X), \tau_1)$ infrabarreled, and so is a Mackey space (Corollary 4.9). But $(T(X), \tau_2)$ is first countable, so it is bornological (Proposition 4.10), hence is infrabarreled, hence is also a Mackey space. That is, *both τ_1 and τ_2 agree with the Mackey topology on $T(X)$, where its dual space is $Y^*/T(X)^\perp$.* In particular, $\tau_1 = \tau_2$. But τ_1 is B-complete (Proposition 5.32) and so is complete (Proposition 5.31), so $T(X)$ is a complete, hence closed (Proposition 1.30) subspace of Y. (If X is a Fréchet space, then so is $T(X)$, making things even simpler.) $\square$

There is one more result along these lines that is suitable for presentation here. It concerns strong topologies. It is difficult to state without replacing Y with $T(X)^-$— but then, *any* closed range theorem can be restated this way.

Proposition 5.39. *Suppose X and Y are Hausdorff locally convex spaces, and suppose X is a Fréchet space and Y is infrabarreled. Suppose $T : X \to Y$ is a continuous linear map for which $T(X)^- = Y$. Then T^* is one-to-one. If $(T^*)^{-1} : T^*(Y^*) \to Y^*$ is strongly bounded, then T has closed range.*

Proof. T^* is one-to-one since T has dense range (Theorem 5.36(a)).

Suppose U is a barrel neighborhood of 0 in X. Then U° is equicontinuous, hence is strongly bounded in X^* [Theorem 4.16(a)], so $U^\circ \cap T^*(Y^*)$ is bounded in $T^*(Y^*)$, so $(T^*)^{-1}(U^\circ \cap T^*(Y^*)) = (T^*)^{-1}(U^\circ)$ is strongly bounded in Y^* by assumption. But $(T^*)^{-1}(U^\circ) = T(U)^\circ$ by Theorem 5.2, so $T(U)^\circ$ is strongly bounded in Y^*. Hence $T(U)^\circ$ is equicontinuous by Theorem 4.16(b), so that $(T(U)^\circ)_\circ$ is a neighborhood of 0 in Y. But $T(U)$ is nonempty, convex, and balanced since U is a barrel, so $(T(U)^\circ)_\circ = T(U)^-$ by the bipolar theorem.

The preceding shows that $T(U)^-$ is a neighborhood of 0 in Y whenever U is a barrel neighborhood of 0 in X, so T is nearly open (Corollary 4.33), and so is onto (Theorem 4.35(b)).   □

Remark. If X and Y are Banach spaces, then the strong topologies on the dual spaces are Banach space topologies, and the preceding theorem shows that if T^* has a strongly closed range, then T has a closed range. [The boundedness of $(T^*)^{-1}$ comes from the open mapping theorem.] This is also true for Fréchet spaces (Theorem 6.1), but that is a bit more involved. Like the result for Banach spaces, it depends on Proposition 5.39.

The final topic here concerns compact linear maps. If X and Y are Hausdorff locally convex spaces, and $T : X \to Y$ is a linear map, then T is **compact** when $T(U)^-$ is compact in Y for some neighborhood U of 0 in X. The next result gives what we need for preparation.

Proposition 5.40. *Suppose X and Y are Hausdorff locally convex spaces over $\mathbb{R}$, and $T : X \to Y$ is a compact linear map. Then T is continuous. If U is a barrel neighborhood of 0 in X for which $T(U)^-$ is compact, and V is a neighborhood of 0 in Y, then there exists a closed subspace E of X such that X/E is finite-dimensional and $T(U \cap E)^- \subset V$.*

Proof. Suppose T is compact, and U is a neighborhood of 0 in X for which $T(U)^-$ is compact. If V is a neighborhood of 0 in Y, then there exists $c > 0$ such that $T(U)^- \subset cV$ since $T(U)^-$ is bounded, so $T(U) \subset cV$, giving $T(c^{-1}U) = c^{-1}T(U) \subset V$ and $T^{-1}(V) \supset c^{-1}U$. This shows that T is continuous (Proposition 1.26(a)).

Now suppose that U is a barrel neighborhood of 0 in X for which $T(U)^-$ is compact, and V is a neighborhood of 0 in Y. Replacing with a smaller neighborhood we may assume that V is a convex, balanced, open neighborhood of 0; it remains to show that there exists a closed subspace E of finite codimension in X for which $T(E \cap U)^- \subset V$.

Suppose not; suppose $T(E \cap U)^- - V \neq \emptyset$ for all closed subspaces $E \subset X$ with $\dim(X/E) < \infty$. If E and F are two closed subspaces of finite codimension in X, then $E \cap F$ is a closed subspace of finite codimension $(E/(E \cap F) \approx (E + F)/F$ is finite dimensional since $(E + F)/F \subset X/F)$ in X, and $T(U \cap E \cap F) \subset T(U \cap E) \cap T(U \cap F)$, giving $T(U \cap E \cap F)^- \subset T(U \cap E)^- \cap T(U \cap F)^-$. It follows that the family of closed sets

$$\{T(E \cap U)^- - V : E \text{ is a closed subspace}$$

$$\text{of } X \text{ with } \dim(X/E) < \infty\}$$

has the finite intersection property in the compact set $T(U)^-$, and so its intersection is nonempty. Suppose $y \in T(E \cap U)^- - V$ for all closed subspaces E of finite codimension in X. This leads to a contradiction, as follows:

First, choose $f \in Y^*$ for which $f(y) \neq 0$ (Corollary 3.17). Set $E = \ker(T^*(f))$, a closed subspace of codimension 1. Note that

$$T(E \cap U) \subset T(E) = T(\ker(T^*(f))) = T(\ker(f \circ T))$$

$$= T(T^{-1}(\ker(f))) \subset \ker(f),$$

and $\ker(f)$ is closed, so $y \notin \ker(f) \supset T(E \cap U)^-$.   □

We can now give the closed range result. It is part of what is commonly called the *Fredholm alternative*.

Theorem 5.41 (Riesz–Leray). *Suppose X is a Hausdorff locally convex space, and suppose $T : X \to X$ is a compact linear transformation. If I denotes the identity map, then $\ker(I - T)$ is finite-dimensional and $(I - T)$ has closed range.*

Proof. Let U_0 be a neighborhood of 0 in X for which $T(U_0)^-$ is compact. U_0 contains a barrel neighborhood U of 0, and $T(U)^- \subset T(U_0)^-$, so $T(U)^-$ is compact as well. If $x \in \ker(I - T)$, then $0 = x - T(x)$, that is $T(x) = x$ so $T(U \cap \ker(I - T)) = U \cap \ker(I - T)$, a closed subset of $T(U)^-$. That is, $U \cap \ker(I-T)$ is a compact neighborhood of 0 in $\ker(I-T)$, so $\ker(I-T)$ is finite-dimensional (Corollary 2.11). Choose a closed subspace E of finite codimension in X for which $T(U \cap E)^- \subset \frac{1}{2}\text{int}(U)$. First of all it suffices to show that $(I-T)(E)$ is closed, since $(I-T)(X)/(I-T)(E)$ is finite dimensional (it is an image of X/E), and so will be closed in $X/(I-T)(E)$ (Corollary 2.10), making $(I-T)(X)$ closed (Theorem 1.23(b)).

To show that $(I - T)(E)$ is closed, start with a calculation. Let p_U denote the Minkowski functional of U; then p_U is a seminorm (Theorem 3.7). Suppose $x \in E$ and $p_U(x) = r$. Then for all $\varepsilon > 0$, $x \in (r + \varepsilon)U$, so that $(r + \varepsilon)^{-1}x \in U \cap E$. Thus $(r + \varepsilon)^{-1}T(x) = T((r + \varepsilon)^{-1}x) \in T(U \cap E) \subset \frac{1}{2}\text{int}U \subset \frac{1}{2}U$, so $T(x) \in \frac{1}{2}(r + \varepsilon)U$, giving $p_U(T(x)) \leq \frac{1}{2}(r + \varepsilon)$. This holds for all $\varepsilon > 0$, so $p_U(T(x)) \leq \frac{1}{2}r = \frac{1}{2}p_U(x)$.

Now suppose $y \in (I - T)(E)^-$. Choose a net $\langle x_\alpha : \alpha \in D \rangle$ in E such that $(I - T)(x_\alpha) = x_\alpha - T(x_\alpha) \to y$. Then $\langle x_\alpha - T(x_\alpha) \rangle$ is a Cauchy net, so there exists α_0 such that

$$\beta, \gamma \succ \alpha_0 \Rightarrow (x_\beta - T(x_\beta)) - (x_\gamma - T(x_\gamma)) \in \frac{1}{2} \text{ int } U.$$

Given $\beta, \gamma \succ \alpha_0$, set $x = x_\beta - x_\gamma$; the above says that $x - T(x) \in \frac{1}{2} \text{ int } U = \frac{1}{2}[0, 1)U$, so $x - T(x) \in rU$ for some $r < \frac{1}{2}$, giving $p_U(x - T(x)) < \frac{1}{2}$. But this means that

$$p_U(x) = p_U(x - T(x) + T(x)) \le p_U(x - T(x)) + p_U(T(x))$$

$$\le p_U(x - T(x)) + \frac{1}{2} p_U(x) < \frac{1}{2} + \frac{1}{2} p_U(x)$$

so that $\frac{1}{2} p_U(x) < \frac{1}{2}$, or $p_U(x) < 1$. Thus $x \in U$. In particular, setting $\gamma = \alpha_0$: $\beta \succ \alpha_0 \Rightarrow x_\beta - x_{\alpha_0} \in U$, so that $x_\beta \in x_{\alpha_0} + U$ and $T(x_\beta) \in T(x_{\alpha_0}) + T(U) \subset T(x_{\alpha_0}) + T(U)^-$, a compact set. By Proposition 1.5, the net $\langle T(x_\beta) \rangle$ on the directed set $\{\beta \in D : \beta \succ \alpha_0\}$ has a cluster point z. The final claim is that $y + z \in E$ and $(I - T)(y + z) = y$. The simplest way to do this is to define a new net.

Let $\mathscr{B}$ be a neighborhood base at 0, and set

$$D' = D \times \mathscr{B}; (\alpha, V) \succ (\beta, W) \text{ when } \alpha \succ \beta \text{ and } V \subset W.$$

D' is directed, and one can define a net on D' as follows. Given $(\alpha, V) \in D'$, choose $\beta \in D$ such that $\beta \succ \alpha$, $\beta \succ \alpha_0$, and $T(x_\beta) \in z + V$, which is possible since z is a cluster point. Let $\beta(\alpha, V)$ be this β. First, note that $\lim_{D'} T(x_{\beta(\alpha,V)}) = z$ by construction. Also, given $W \in \mathscr{B}$, there exists α_1 such that $\alpha \succ \alpha_1 \Rightarrow x_\alpha - T(x_\alpha) \in y + W$. But now if $(\alpha, V) \succ (\alpha_1, W)$, then $\beta(\alpha, V) \succ \alpha \succ \alpha_1$, so that $x_{\beta(\alpha,V)} - T(x_{\beta(\alpha,V)}) \in y + W$ as well. This shows that $\lim_{D'} (I - T)(x_{\beta(\alpha,V)}) = y$. We therefore get that

$$\lim_{D'} x_{\beta(\alpha,V)} = \lim_{D'} (x_{\beta(\alpha,V)} - T(x_{\beta(\alpha,V)})$$
$$+ \lim_{D'} T(x_{\beta(\alpha,V)}) = y + z \in E$$

since E is closed, and

$$T(y + z) = T(\lim_{D'} x_{\beta(\alpha,V)})$$
$$= \lim_{D'} T(x_{\beta(\alpha,V)}) = z$$

since T is continuous. Hence $(I - T)(y+z) = y+z-T(y+z) = y+z-z = y$. $\square$

The construction at the end of the proof is an example of a *subnet*; notice that it is much more complicated than the construction of a subsequence, since the directed set must be allowed to change.

It would be nice if we could conclude that T^* is (strongly) compact when T is compact, but this is not so in general. However, there is a topology one can place on the dual spaces, called the *Arens topology*, under which the adjoint of a compact operator is compact. This allows one to conclude that $X/(I - T)(X)$ is also finite-dimensional. All this is covered in the exercises. Somewhat deeper is the fact that $X/(I - T)(X)$ and $\ker(I - T)$ have the same dimension. That also appears in the exercises, with the following result providing the starting point. It also gives a weaker condition under which $\ker(I - T)$ will be finite-dimensional, but compactness for T also appears to be necessary to force $I - T$ to have closed range.

Proposition 5.42. *Suppose X is a Hausdorff locally convex space, $T : X \to X$ is a continuous linear transformation, and U is a barrel neighborhood of 0 subject to:*

(α) *$T(U)$ does not contain a nontrivial subspace of X, and*
(β) *$T(U)$ is covered by N translates of $\frac{1}{2}\mathrm{int}(U)$; that is there exists $w_1, \ldots, w_N \in X$ for which*

$$T(U) \subset \bigcup_{j=1}^{N} \left[w_j + \frac{1}{2}\mathrm{int}(U) \right].$$

Consider the chain of subspaces $K_j = \ker(I - T)^j$:

$$\{0\} \subset K_1 \subset K_2 \subset \cdots.$$

Then:

(a) The chain stabilizes beyond $j = N$: $K_N = K_{N+1} = \cdots$;
(b) Every K_j has dimension $\leq N$; and
(c) $\dim K_1 = \dim \ker(I - T) \leq \dim(X/(I - T)(X))$.

In the proof, the following lemma will be used several times:

Lemma 5.43. *Assume X, T, and U are as in Proposition 5.42. Suppose*

$$0 = M_0 \subsetneqq M_1 \subsetneqq M_2 \subsetneqq \cdots \subsetneqq M_n$$

is a chain of finite dimensional subspaces of X for which $(I - T)M_k \subset M_{k-1}$ when $k \geq 1$. Then $n \leq N$.

Preliminary Observation: Let p_U denote the Minkowski functional associated with U. By Theorem 3.7, p_U is a continuous seminorm since U is a neighborhood of 0 and is convex and balanced. Letting I_x be as in the definition of p_U, then

$$I_x = \{t > 0 : x \in tU\} = \{t > 0 : t^{-1}x \in U\},$$

a relatively closed subset of the interval $(0, \infty)$ since U is closed. Thus (unless $p_U(x) = 0$), $I_x = [p_U(x), \infty)$, so $x \in U \Leftrightarrow 1 \in [p_U(x), \infty) \Leftrightarrow p_U(x) \leq 1$. That is:

$$U = \{x \in X : p_U(x) \leq 1\}. \tag{$*$}$$

(This has been noted before, using sequences.)

Also, $\text{int}(U) = [0, 1)U$ (Theorem 2.15), so if $x \in \text{int}(U)$, then $x = ty$ for some $y \in U$ and $t \in [0, 1)$, giving $p_U(x) = p_U(ty) = tp_U(y) \leq t \cdot 1 < 1$. On the other hand, if $p_U(x) < 1$, choose t for which $p_U(x) < t < 1$. Then $p_U(t^{-1}x) = t^{-1}p_U(x) < t^{-1}t = 1$, so $t^{-1}x \in U$ and $x = t \cdot t^{-1}x \in [0, 1)U = \text{int}(U)$. Hence

$$\text{int}(U) = \{x \in X : p_U(x) < 1\}. \tag{$**$}$$

Proof of Lemma 5.43. The first thing to note is that p_U is a norm on each M_k. This is by induction on k, and is trivial when $k = 0$. As for $k \to k + 1$, suppose $x \in M_{k+1}$ and $p_U(x) = 0$. Letting $\mathbb{F}$ denote the scalar field ($\mathbb{R}$ or $\mathbb{C}$), if $c \in \mathbb{F}$, then $p_U(cx) = 0$ as well, so $cx \in U$ for all $c \in F$. Thus $cT(x) = T(cx) \in T(U)$ for all $c \in F$, that is $\mathbb{F} \cdot T(x)$ is a subspace of X contained in $T(U)$. By assumption, this must be trivial, so $T(x) = 0$. Hence $x = x - T(x) = (I - T)(x) \in M_k$, so that $x = 0$ by the induction hypothesis (p_U *is a seminorm on* M_k).

Next, there is only one way to make a finite-dimensional space into a Hausdorff locally convex space (Proposition 2.9), and on each M_k, the norm topology from p_U does that, so p_U gives the induced topology on each M_k. Also, $U \cap M_k$ is not contained in M_{k-1} ($U \cap M_k$ is absorbent in M_k), and $2U \cap M_k$ is compact. By $(*)$, if $x \in U \cap M_k$ and $y \notin 2U$, then $p_U(x) \leq 1$ while $p_U(y) \geq 2$. Since $p_U(y) = p_U(x + (y - x)) \leq p_U(x) + p_U(y - x)$: $p_U(y - x) \geq 1$.

For $k = 1, \ldots, n$, choose any $y_k \in U \cap M_k - M_{k-1}$. As a function on M_{k-1},

$$f_k(z) = p_U(y_k - z)$$

has a value ≤ 1 at $z = 0 \in U \cap M_{k-1}$, while it has values ≥ 1 for $z \in M_{k-1} - 2U$, so the minimum of f_k on the compact set $2U \cap M_{k-1}$ is a minimum on M_{k-1}. Let z_k be a point where this minimum is achieved, with $t_k = p_U(y_k - z_k)$. Set $x_k = t_k^{-1}(y_k - z_k)$. Observe the following:

(i) $p_U(x_k) = p_U(t_k^{-1}(y_k - z_k)) = t_k^{-1}p_U(y_k - z_k) = 1$, so $x_k \in U$ by $(*)$.
(ii) If $z \in M_{k-1}$, then

$$p_U(x_k - z) = p_U(t_k^{-1}(y_k - z_k) - z)$$
$$= t_k^{-1}p_U(y_k - (z_k + t_kz)) \geq t_k^{-1}t_k = 1.$$

since $z_k + t_kz \in M_{k-1}$. Hence
(iii) (Trick Alert!) If $k > j$, then $x_j - T(x_j) \in M_{j-1} \subset M_{k-1}$ and $x_j \in M_j \subset M_{k-1}$, so

$$T(x_k) - T(x_j) = x_k - \underbrace{(x_k - T(x_k)) + x_j - (x_j - T(x_j))}_{\text{in } M_{k-1}}, \text{ and}$$

$$p_U(T(x_k) - T(x_j)) = p_U(x_k - (x_k - T(x_k)) + x_j - (x_j - T(x_j)))_{\geq 1}.$$

Now $T(x_1), \ldots, T(x_n) \in T(U)$, which is covered by the sets $w_l + \frac{1}{2}\text{int}(U)$. If both $T(x_j)$ and $T(x_k)$ belong to $w_l + \frac{1}{2}\text{int}(U)$, then $T(x_k) - w_l \in \frac{1}{2}\text{int}(U)$, so $p_U(T(x_k) - w_l) < \frac{1}{2}$. Similarly, $p_U(T(x_j) - w_l) < \frac{1}{2}$, so

$$p_U(T(x_k) - T(x_j)) \leq p_U(T(x_k) - w_l + w_l - T(x_j))$$
$$\leq p_U(T(x_k) - w_l) + p_U(w_l - T(x_j)) < 1.$$

Since this cannot happen: The points $T(x_1), \ldots, T(x_n)$ must belong to distinct sets $w_1 + \frac{1}{2}\text{int}(U), \ldots, w_N + \frac{1}{2}\text{int}(U)$. By the pigeon hole principle, $n \leq N$. $\square$

Proof of Proposition 5.42: This is done using a series of steps.

Step 1:   dim $K_1 \leq N$. Suppose $v_1, \ldots, v_n$ is a finite, linearly independent subset of K_1. Set $M_k = \text{span}\{v_1, \ldots, v_k\}$. Since $(I - T)M_k = \{0\}$, these spaces satisfy the hypotheses of Lemma 5.43, so $n \leq N$. Since N is an upper bound for any finite linearly independent subset of K_1, and K_1 does have a basis (which, if infinite, will have arbitrarily large finite subsets), K_1 must be finite dimensional, with dimension $\leq N$.

Step 2: $\dim(K_{j+1}/K_j) \leq \dim(K_j/K_{j-1})$. Consider the composite map:

$$K_{j+1} \xrightarrow{(I-T)} K_j \xrightarrow{\pi} K_j/K_{j-1}.$$

The kernel is

$$\{x \in K_{j+1} : (I-T)(x) \in K_{j-1}\} = \{x \in K_{j+1} : (I-T)^{j-1}(I-T)(x) = 0\} = K_j,$$

so $\dim(K_{j+1}/K_j)$ equals the dimension of the image of the composite, which (as a subspace) has dimension $\leq \dim(K_j/K_{j-1})$.

Step 3:   Every K_j is finite-dimensional. Induction on j. Step 1 gives the $j = 1$ case, while Step 2 provides the induction step.

Proof for part (a): Set $M_j = K_j$, now known to be finite-dimensional. Once $K_j = K_{j-1}$, you get $K_{j+1} = K_j$ by Step 2, so it stabilizes beyond some n, with $K_{n-1} \neq K_n = K_{n+1} \cdots$ (unless all $K_j = \{0\}$, in which case Proposition 5.42 is trivial). By Lemma 5.43, $n \leq N$.

Now set $K = K_N = K_{N+1} = \cdots$.

Step 4: $\dim(K) \leq N$ (proving part (b)). Start with a basis of $K_1 : v_1, \ldots, v_l$. Extend it to a basis of $K_2 : v_{l+1}, \ldots v_{l'}$. Etc. As in Step 1, set $M_k = \text{span}\{v_1, \ldots, v_k\}$. Since we do not climb up to K_{j+1} until we have a basis of K_j, this satisfies the hypotheses of Lemma 5.43. Since the basis has $\dim K$ entries, $\dim K \leq N$. [Note that what we really need here to get the spaces M_k is that, relative to the ordered basis $(v_1, \ldots, v_k)$, the matrix of $I - T$ should be in upper triangular form, with zeroes on the diagonal.]

Step 5: There is a subspace Y of K, with $\dim Y = \dim K_1$, for which $Y \cap (I - T)(K) = \{0\}$. This is pure linear algebra. As a map from K to itself, $I - T$ has kernel K_1, so it has rank $\dim K - \dim K_1$ by the rank-nullity theorem. Now choose Y to be a subspace of K which is complementary to $(I - T)(K)$, so that $K = Y \oplus (I - T)(K)$.

Step 6: $Y \cap (I - T)(X) = \{0\}$. This is where part (a) is needed, and resembles arguments from the study of commutative Noetherian rings. Suppose $x \in Y \cap (I - T)(X)$. Since $x \in (I - T)(X)$, we can write $x = (I - T)(y)$ for some $y \in X$. Since $x \in Y \subset K = K_N$:

$$(I - T)^{N+1}(y) = (I - T)^N(I - T)(y) = (I - T)^N(x) = 0.$$

Hence $y \in K_{N+1} = K_N$, so that $x = (I - T)(y) \in (I - T)(K)$. Combining: $x \in Y \cap (I - T)(K) = \{0\}$.

Part (c) now follows:

$$\dim(X/(I - T)(X)) \geq \dim((Y + (I - T)(X))/(I - T)(X))$$

$$= \dim(Y/(Y \cap (I - T)(X))) = \dim(Y) = \dim K_1.$$

□

A final note. When T is actually compact, with $T(U)^-$ being a compact subset of X, then $T(U)$ is bounded, and so cannot contain a nontrivial subspace of X. Also, since $\dim \ker(I - T) = \dim(X/(I - T)(X))$ by Exercise 25 of this chapter, it follows from the proof above that $X = Y \oplus (I - T)(X)$. Also, the bound "$n \leq N$" in Lemma 5.43 can be improved on. When the base field is $\mathbb{R}$, one can use the set $\{\pm x_1, \ldots, \pm x_n\}$ to get $2n \leq N$. When the base field is $\mathbb{C}$, one can get $6n \leq N$ (Do you see why?). One suspects, for measure theoretic reasons, that $n \leq$ constant $\cdot \log(N)$, but complicating matters is the fact that the points w_l need not belong to K. If you are inclined toward combinatorial geometry, have fun with it.

Exercises

1. The following is part of Theorem 5.2.

   If X and Y are locally convex spaces, and $T \in \mathscr{L}_c(X, Y)$, then T^* is continuous when X^* and Y^* are equipped with their weak-* topologies.

Expand on this. Suppose X is also Hausdorff and infrabarreled. Show that $S \in \mathscr{L}_c(Y^*, X^*)$ is an adjoint map (i.e. $S = T^*$ for some $T \in \mathscr{L}_c(X, Y)$) if and only if S is weak-* continuous.

2. Examine the proof of Theorem 5.10, and isolate the proof of the following: If X is a Hausdorff locally convex space, then X is infrabarreled if and only if all strongly bounded subsets of X^* are equicontinuous.

3. Suppose X and Y are Hausdorff locally convex spaces. Show that $T \mapsto T^*$ from $\mathscr{L}_c(X, Y)$ to $\mathscr{L}_c(Y^*, X^*)$ is bounded.

4. Suppose X and Y are Banach spaces. Show that $T \mapsto T^*$ from $\mathscr{L}_c(X, Y)$ to $\mathscr{L}_c(Y^*, X^*)$ is an isometry.

5. The first part of the proof of Proposition 5.32 is a proof that the map $(\pi^*)_0$ of Theorem 5.7 is a homeomorphism of $((X/M)^*$, weak-* topology with $(M^\perp$, topology induced from weak-* topology of X^*). The point of this problem is to prove the analogous result for the map $(\iota^*)_0$ of Theorem 5.6 when M is closed. For this problem, X is a locally convex space, and M is a closed subspace. ι, ι^*, and $(\iota^*)_0$ are as in Theorem 5.6.

(a) Show that ι^* and $(\iota^*)_0$ are continuous when weak-* topologies are used.

(b) Show that a convex, balanced, absorbent subset of $\mathbb{R}^n$ or $\mathbb{C}^n$ is a neighborhood of 0. (Intersect with a ball, then use Proposition 2.9.)

(c) Show that if V is a convex, balanced subset of X^* such that $V = [0, 1)V$, and if E is a subspace of X^* which is contained in V, then V is a union of cosets of E.

(d) Show that if V is a convex, balanced, absorbent subset of X^*, then V is a weak-* neighborhood of 0 if and only if V contains a weak-* closed subspace E of X^* for which X^*/E is finite-dimensional. (*Hint:* You can look at $[0, 1)V$, and use part (c). Proposition 2.9 also helps.)

(e) Suppose E and F are two weak-* closed subspaces of X^*, and suppose X^*/E is finite-dimensional. Show that $E + F$ is weak-* closed, with $E + F = ((E_\perp) \cap (F_\perp))^\perp$. (Exercise 18 from Chap. 1 will help here, as will Proposition 2.9.)

(f) Show that ι^* and $(\iota^*)_0$ are open maps when weak-* topologies are used. (Use $F = M^\perp$ in part (e); note that $F_\perp = M$ since M is closed.)

6. Suppose $\langle X_i : i \in \mathscr{I} \rangle$ is an infinite family of Hausdorff locally convex spaces. Check that the product $\prod X_i$ is a Hausdorff locally convex space. (Theorem 2.1 does most of the work.) Show that algebraically, $(\prod X_i)^* \approx \Sigma \oplus X_i^*$. Also, show that if B_i is bounded in X_i for all i, then $\prod B_i$ is bounded in $\prod X_i$, and every bounded subset of $\prod X_i$ is contained in such a product. Finally, use this to show that the strong topology on $(\prod X_i)^* \approx \Sigma \oplus X_i^*$ is the *box topology*, where a neighborhood base at 0 consists of sets of the form $(\Sigma \oplus X_i^*) \cap \prod U_i$, where the sets U_i vary over neighborhood bases at 0 for each X_i.

7. Suppose X and Y are two Hausdorff locally convex spaces, so that $(X \times Y)^* \approx X^* \times Y^*$ and $(X \times Y)^{**} \approx X^{**} \times Y^{**}$. Show that $J_{X \times Y}$ corresponds to $J_X \times J_Y$ under these identifications. Use this to show that $X \times Y$ is infrabarreled if and only if both X and Y are infrabarreled.

8. Show that a normed linear space that is also a Montel space is finite-dimensional.
9. Show that Example I of Sect. 3.7 is a Montel space.
10. Show that Example I of Sect. 3.8 is a Montel space.
11. Show that, in the proof of Proposition 5.22,

$$T^{-1}\left(\frac{1}{3}U\right) \supset \bigcap_{\alpha \in D} T_\alpha^{-1}\left(\frac{1}{3}U\right),$$

given simply that $T(x) = \lim T_\alpha(x)$ always exists. In particular, just using condition (b), the pointwise limit T (which is linear by Proposition 4.19) is continuous.
12. Suppose X is a Hausdorff locally convex space. Show that X is a Montel space if, and only if, X is quasicomplete and infrabarreled, and bounded sets are (originally) precompact.
13. (Compare with previous problem) Suppose X is a Hausdorff locally convex space.

   (a) Show that bounded subsets of X are weakly precompact.
   (b) Show that X is semireflexive if, and only if, X is weakly quasicomplete.

   Suggestion for (a): If $f_1, \ldots, f_n \in X^*$, look at the image of a bounded set in $\mathbb{F}^n$ (using the f_j as coordinate functions), where $\mathbb{F}$ is the base field. This image is totally bounded in $\mathbb{F}^n$.
14. Suppose X is a Banach space, and $|||?|||_*$ is a norm on X^* that is equivalent to the operator norm on X^*. Show that there exists a norm on X which is equivalent to its original norm, and for which $|||?|||_*$ is its operator norm.
15. Suppose μ is Lebesgue measure on $[0, 1]$. Show that the closed unit ball in $L^1(\mu)$ has no extreme points. Use this, the Krein–Milman theorem, and Exercise 14, to show that $L^1(\mu)$ is not topologically isomorphic to the dual space of *any* Banach space.
16. Compute the extreme points of ℓ^1. *Note:* ℓ^1 "is" the dual space of c_0, the subspace of ℓ^∞ consisting of sequences which tend to zero.
17. Suppose X is a compact Hausdorff space, and let $C(X)$ denote the Banach space of continuous, real valued functions on X. Show that $f \in C(X)$ is an extreme point of the closed unit ball in $C(X)$ if and only if $f(X) \subset \{\pm 1\}$. Use this to show that $C(X)$ cannot be reflexive unless X is totally disconnected. (Note: Once $C(X)^*$ is identified, it is not hard to show that $C(X)$ is reflexive if and only if X is finite.)
18. Suppose X is a Hausdorff locally convex space, and M is a closed subspace. Prove:

   (a) If X is B_r-complete, then so is M.
   (b) If X is B complete, then so is M.

   (Exercise 5 will help here.)
19. Show that the strong dual of a reflexive Fréchet space is B-complete.

20. Suppose X is a barreled, Hausdorff, locally convex space, and suppose X is not B_r-complete. This problem consists of constructing a discontinuous linear map from a barreled space to X which has closed graph. Let M denote a weak-* dense, proper subspace of X^* having the property that $M \cap U^\circ$ is weak-* closed for all barrel neighborhoods U of 0 in X. Let τ denote the topology of X. Set

$$\mathscr{B}_0 = \{(M \cap U^\circ)_\circ : U \text{ is a barrel}$$

neighborhood of 0 in $X\}$.

Explain why statements (a)–(n) are true. (Most take no more than a sentence or two.)

(a) $\mathscr{B}_0$ is a base at 0 for a Hausdorff locally convex topology τ_0 on X.
(b) The dual space of (X, τ_0) is M.
(c) $\tau_0 \subsetneqq \tau$.
    Let T denote the identity map from the vector space X to itself, considered as a map from (X, τ_0) to (X, τ). Its graph is the diagonal in $X \times X$.
(d) $\Gamma(T)$ is $\tau_0 \times \tau_0$-closed.
(e) $\Gamma(T)$ is $\tau_0 \times \tau$-closed.
(f) $T : (X, \tau_0) \to (X, \tau)$ is not continuous.
    Let U denote a barrel in (X, τ_0). Set $D = U^\circ$ and $E = M \cap U^\circ$
(g) U is a barrel in (X, τ).
(h) E is the polar of U as a subset of (X, τ_0).
(i) $E_\circ = U$ and $D_\circ = U$.
(j) E and D are both weak-* closed, convex, balanced, and nonempty in X^*.
(k) $D = E$.
(l) U is a τ_0-neighborhood of 0 in X.
    Letting U vary
(m) (X, τ_0) is barreled.
    Note: (e), (f), and (m) complete the discussion, except for:
(n) Fréchet spaces are B_r-complete.

21. (a) Suppose X is a Hausdorff locally convex space. Reverse the argument in Proposition 5.32 to show that X is B-complete provided X/Y is B_r-complete whenever Y is a closed subspace of X. (Given M, set $Y = M_\perp$.)
    (b) Show that Fréchet spaces are B-complete.
22. Suppose X is an infinite-dimensional normed space. Show that the weak topology on X is not first countable, hence is not metrizable. *Hint:* Look closely at the second half of the proof of Proposition 5.38.
23. Suppose X is a Hausdorff locally convex space. The *Arens topology* on X^* is defined by the base

$$\{K^\circ : K \text{ is originally compact}$$

and convex in $X\}$.

verify that this is a locally convex topology on X^*. (The discussion of the Mackey topology in Sect. 3.6 will help here.) Show also that if E is an equicontinuous subset of X^*, then the Arens topology on E coincides with the weak-* topology. (Use Proposition 5.22.) Hence show that if U is a barrel neighborhood of 0 in X, then U° is compact in the Arens topology. Finally, show that if X^* is given the Arens topology, then every continuous linear functional on X^* is evaluation at a point of X.

24. Suppose X and Y are Hausdorff locally convex spaces, and $T : X \to Y$ is a compact linear map. Show that T^* is compact when X^* and Y^* are equipped with the Arens topology. (Exercise 23)

25. Suppose X is a Hausdorff locally convex space, and $T : X \to X$ is a compact linear map. Show that $X/(I - T)(X)$ is finite-dimensional, with the same dimension as $\dim \ker(I - T)$. (I = identity map. Use Exercise 23 and Proposition 5.42.)

The next six exercises are concerned with Fredholm operators, although some preliminary results are included. If X and Y are two Fréchet spaces or two LF-spaces, then a continuous linear map $S : X \to Y$ is called a "Fredholm operator" if S has closed range, and both $\ker(S)$ and $Y/S(X)$ are finite-dimensional. The *index* of S is defined as $\dim \ker S - \dim(Y/S(X))$. It is important in a number of applications. From Theorem 5.41 and Exercise 25, if T is compact and $X = Y$, then $I - T$ is a Fredholm operator of index zero. Exercise 31 is the ultimate objective: If S is a Fredholm operator and $T : X \to Y$ is compact, then $S - T$ is also Fredholm, with the same index as S. This result is sometimes referred to as the "homotopy theory of Fredholm operators."

26. Suppose X, Y, and Z are three Hausdorff locally convex spaces, and suppose $S : Y \to Z$ and $T : X \to Y$ are two continuous linear maps, one of which is compact. Show that $ST : X \to Z$ is compact. (*Note:* There is a slight trick involved if the compact map is T.)

27. Suppose X and Y are two Hausdorff locally convex spaces. Show that the set of compact linear maps from X to Y constitutes a subspace of $\mathscr{L}_c(X, Y)$, which includes all continuous linear maps of finite rank. (*Hint:* The latter factor through $\mathbb{F}^n$.)

28. Suppose X and Y are two Hausdorff locally convex spaces, $\Phi : X \to Y$ is a linear homeomorphism, and $T : X \to Y$ is a compact linear map. Show that $\Phi - T$ has closed range, with $\dim(\ker(\Phi - T)) = \dim(Y/(\Phi - T)(X)) < \infty$.

29. Suppose $X = \bigcup X_n$ is an LF-space, and Y is a Fréchet space. Show that $X \times Y = \bigcup(X_n \times Y)$ is an LF-space. That is, show that the product topology on $X \times Y$ coincides with the LF-topology associated with writing $X \times Y = \bigcup(X_n \times Y)$. (Use Lemma 5.8.) (*Note:* A similar result, with a similar proof shows that $X \times Y$ is an LF-space when both X and Y are LF-spaces.)

30. Suppose X and Y are either two Fréchet spaces or are two LF-spaces, and $S : X \to Y$ is a Fredholm operator of index 0.

(a) Show that there is a continuous linear map $T_0 : X \to Y$ of finite rank for which $\Phi = S + T_0$ is a linear homeomorphism. [*Hint:* Have T_0 map $\ker(S)$ onto a subspace complementary to $S(X)$, and use Theorem 4.35, Theorem 4.37, or Exercise 31 from Chap. 4 to cover Φ^{-1}.]

(b) Show that if $T : X \to Y$ is compact, then $S - T$ is Fredholm with index 0. (Do you really need a hint? $S - T = \Phi - (T + T_0)$.)

31. Suppose X and Y are either two Fréchet spaces or two LF-spaces, and suppose $S : X \to Y$ is a Fredholm operator and $T : X \to Y$ is compact. Show that $S - T$ is Fredholm, with the same index as S.
    Hint/Trick: If the index of S is positive, replace Y with $Y \times \mathbb{F}^r$. If it is negative, ...(This is why Exercise 29 is grouped here.)

One more problem, concerning completeness of quotients.

32. Suppose X is a locally compact Hausdorff space. Let $C(X)$ denote the space of continuous, real-valued functions on X. Using Definition 3.32, topologize $C(X)$ using the directed family of seminorms

$$p_K(f) = \max\{|f(K)| : K \text{ compact in } X\}.$$

Suppose Y is a closed subset of X, and consider the map $f \mapsto f|_Y$, with (closed) kernel M and range $\mathscr{A} \subset C(Y)$. *Note:* Y is also locally compact. Topologize $C(Y)$ in the same way, using compact $K \subset Y$.

(a) Show that $C(X)$ is complete.

(b) Show that $\mathscr{A}$ is dense in $C(Y)$ using the Stone–Weierstrass theorem on each compact $K \subset Y$.

(c) If K is compact in X, let $\hat{K}$ denote $(K - Y) \bigcup \{\infty\}$, the one point compactification of $K - Y$. (*Note:* As a topological space, $K - Y$ is locally compact, so $\hat{K}$ is a compact Hausdorff space.) If $g \in C(K)$ (or $g \in C(X)$) and $g|_{K \cap Y} \equiv 0$, set

$$\hat{g}(x) = \left\{ \begin{array}{ll} g(x) & \text{if } x \in K - Y \\ 0 & \text{if } x = \infty \end{array} \right\} \text{ for } x \in \hat{K}.$$

Using the Stone-Weierstrass theorem on $\hat{K}$, show that $\{\hat{g} : g \in M\}$ is dense in $\{h \in C(\hat{K}) : h(\infty) = 0\}$. Use this to prove the following: If $f \in C(X)$, $p_{K \cap Y}(f) = m$, and $\varepsilon > 0$, then one can set

$$A = \{x \in K : |f(x)| \geq m + \varepsilon\},$$

and letting χ be a Urysohn function on K which is 0 on $K \cap Y$ and 1 on A, there exists $h \in M$ such that $\max |(\widehat{\chi f} - \hat{h})(\hat{K})| < \varepsilon$. Show that for this h, $p_K(f - h) < m + 2\varepsilon$. Finally, use this to show that the subspace topology on $\mathscr{A} \subset C(Y)$ coincides with the quotient topology on $C(X)/M \approx \mathscr{A}$.

(d) Now suppose that X is not normal. (Such spaces do exist; see Proposition A.21 in Appendix A.) Let A and B denote two closed subsets that cannot be separated. Set $Y = A \cup B$. Show that the function that is 1 on A an 0 on B is continuous on Y but does not belong to $\mathscr{A}$. Hence show that $C(X)/M$ is not complete.

(e) ("Mathematics Made Difficult"—Linderholm [25]) Use (a)–(c) and the idea behind (d) to show that a σ-compact, locally compact Hausdorff space is normal.

Chapter 6
Duals of Fréchet Spaces

6.1 Overview Plus

A good title for this chapter might have been "Weird Countability." The point is that, while "countability" applies to a Fréchet space X in basically one way (it is first countable), it affects X^* in some rather strange ways.

There are four major theorems to be proven in this chapter. Each uses its own twist on countability. Letting X denote a Fréchet space, they read as follows:

> 1. *(Krein–Smulian I) A convex subset E of X^* is* weak-* *closed provided $E \cap U^\circ$ is* weak-* *closed for all barrel neighborhoods U of 0 in X.*

Note that the fact that Fréchet spaces are B-complete follows immediately.

> 2. X^{**} *is a Fréchet space.*

The fact that X^{**} is first countable is easy; the problem is completeness, since X^* need not be bornological. The conditions from Definition 4.1 are directly related, however.

> 3. *The following are equivalent:*
>   (i) X^* *is barreled.*
>   (ii) X^* *is infrabarreled.*
>   (iii) X^* *is bornological.*

Proving (ii) $\Rightarrow$ (iii) is the hard part.

Finally, there is the leftover from Sect. 5.7. Surprisingly, it is feasible to do this now.

Theorem 6.1. *Suppose X and Y are Fréchet spaces, $T \in \mathscr{L}_c(X, Y)$, and T^* has strongly closed range. Then T has closed range.*

M.S. Osborne, *Locally Convex Spaces*, Graduate Texts in Mathematics 269,
DOI 10.1007/978-3-319-02045-7_6, © Springer International Publishing Switzerland 2014

Proof. First, replace Y with $T(X)^-$, which is also a Fréchet space. We have (dually):

$$\left[\begin{array}{l} X \xrightarrow{T} T(X)^- \hookrightarrow Y \\ X^* \xleftarrow{\text{new } T^*} Y^*/T(X)^\perp \underset{\text{onto}}{\longleftarrow} Y^*. \end{array}\right.$$

The composite along the bottom is the old T^*, and its range is the same as that of the new T^*. The whole point is to get to the situation of Proposition 5.39. So having made this replacement, it suffices to show that $(T^*)^{-1} : T^*(Y^*) \to Y^*$ is strongly bounded.

Suppose B is a strongly bounded subset of $T^*(Y^*)$. B is equicontinuous [Theorem 4.16(b)], so there exists a barrel neighborhood U of 0 such that $B_\circ \supset U$, giving $B \subset U^\circ$ (Proposition 3.19(e)). Set $D = T^*(Y^*) \bigcap U^\circ$, a strongly closed, convex, balanced, strongly bounded, nonempty subset of X^*. (U° is strongly bounded since it is equicontinuous by definition.) In accordance with Proposition 3.30, form the normed space (X_D^*, p_D); this is a Banach space since X^* (and hence D) is complete (Corollary 4.22).

Now for the "weird countability." Let $V_1 \supset V_2 \supset \cdots$ be a neighborhood base for the topology of Y at 0. If $f \in Y^*$, then $\{f\}_\circ$ is a neighborhood of 0, so $V_n \subset \{f\}_\circ$ for some n, so that $f \in V_n^\circ$. That is, $Y^* = \bigcup V_n^\circ$. Thus $T^*(Y^*) = T^*(\bigcup V_n^\circ) = \bigcup T^*(V_n^\circ)$. Each V_n° is weak-* compact, so each $T^*(V_n^\circ)$ is weak-* compact, hence is weak-* closed, and so is strongly closed. It follows that $T^*(V_n^\circ) \bigcap X_D^*$ is p_D-closed in the Banach space (X_D^*, p_D). (Proposition 3.30). But $D \subset T^*(Y^*)$, a subspace of X^*, so $X_D^* = $ domain of $p_D = \bigcup nD \subset T^*(Y^*)$, giving that $X_D^* = \bigcup(T^*(V_n^\circ) \bigcap X_D^*)$. By Baire category, some $T^*(V_n^\circ) \bigcap X_D^*$ has nonempty interior in X_D^*, and that interior is convex and balanced (since $T^*(V_n^\circ) \bigcap X_D^*$ is) and so contains 0. Hence, for some $\varepsilon > 0 : \varepsilon D \subset T^*(V_n^\circ) \bigcap X_D^*$.

This does it: $D \subset \varepsilon^{-1} T^*(V_n^\circ) = T^*(\varepsilon^{-1} V_n^\circ)$, giving:

$$B = T^*(Y^*) \bigcap B \subset T^*(Y^*) \bigcap U^\circ = D \subset T^*(\varepsilon^{-1} V_n^\circ), \text{ that is}$$

$$(T^*)^{-1}(B) \subset \varepsilon^{-1} V_n^\circ = (\varepsilon V_n)^\circ,$$

an equicontinuous, hence strongly bounded (Theorem 4.16(a)) set.   □

Corollary 6.2. *Suppose X and Y are Fréchet spaces, and $T \in \mathscr{L}_c(X, Y)$. The following are equivalent:*

(i) T has closed range.

(ii) T^* has weak- closed range.*

(iii) T^* has strongly closed range.

Proof. (iii) $\Rightarrow$ (i) is Theorem 6.1. (i) $\Rightarrow$ (ii) is Proposition 5.37. (ii) $\Rightarrow$ (iii) is trivial.

□

6.2   Krein–Smulian I

The name "Krein–Smulian theorem" is generally applied to two results. The one here concerns weak-* closed subsets of the dual of a Fréchet space, as described in Sect. 6.1. The other appears in Appendix C.

Since we are concerned with looking at sets of the form $C \cap U^\circ$ in X^*, it is useful to use this to define a topology.

Definition 6.3. Suppose X is a locally convex space, and $C \subset X^*$. C is **almost weak-* closed** (resp. **almost weak-* open**) if $C \cap U^\circ$ is weak-* closed (resp. relatively weak-* open in U°) for all barrel neighborhoods U of 0 in X. τ_a denotes the class of almost weak-* open subsets of X^*.

Before going on, two more quick generalities: Define τ_K to be the locally convex topology on X^* with a base at 0 that consists of all K°, where K is compact in X.

Let τ_N denote the locally convex topology on X^* whose base at 0 consists of all S°, where $S = \{x_n : n \in \mathbb{N}\}$ for a sequence x_n such that $\lim x_n = 0$. These both give locally convex topologies on X^* in the usual way. (If $S = \{x_n\}$ and $T = \{y_n\}$, then $S \bigcup T = \{z_n\}$, where $z_{2n} = x_n$ and $z_{2n+1} = y_n$.)

The next result is basic.

Lemma 6.4. *Suppose X is a locally convex space. Let τ^* denote the weak-* topology on X^*. Then:*

(a) *C is almost weak-* open if, and only if, $C \cap E$ is relatively weak-* open in E for all equicontinuous sets E.*

(b) *τ_a is a translation invariant topology.*

(c) *$\tau^* \subset \tau_N \subset \tau_K \subset \tau_a$.*

Proof. (a) If C is almost weak-* open, and E is equicontinuous, then there exists a barrel neighborhood U of 0 such that $U^\circ \supset E$. Hence $C \cap E = (C \cap U^\circ) \cap E$ is relatively weak-* open in E since $C \cap U^\circ$ is relatively weak-* open in U°. Conversely, if $C \cap E$ is relatively weak-* open in E for all equicontinuous E, then $C \cap U^\circ$ is relatively weak-* open in U° because U° is equicontinuous.

(b) If C is τ_a-open, and $f \in X^*$, then for all equicontinuous E:

$$(f + C) \cap E = f + (C \cap (E - f))$$

is weak-* open in $E - f$ since the weak-* topology is translation invariant and $E - f$ is equicontinuous. That τ_a is a topology comes from

$$\left[\begin{array}{l} (\bigcup C_i) \cap E = \bigcup (C_i \cap E) \text{ and} \\ (C \cap C') \cap E = (C \cap E) \cap (C' \cap E). \end{array} \right.$$

(c) $\tau^* \subset \tau_N$ because each τ^*-neighborhood of 0 is a τ_N-neighborhood: If $F = \{x_1, \ldots, x_n\}$ is finite, set $x_k = 0$ for $k > n$. Then $\lim x_k = 0$, and $F^\circ = (F \cup \{0\})^\circ$ is a τ_N-neighborhood of 0. $\tau_N \subset \tau_K$ as follows. If $S = \{x_n\}$ for a sequence x_n such that $\lim x_n = 0$, then $\{x_n\} \cup \{0\}$ is compact (standard topology), so $S^\circ = (S \cup \{0\})^\circ$ is a standard neighborhood of 0 for τ_K. Finally, suppose C is τ_K-closed. If U is a barrel neighborhood of 0, and f belongs to the weak-* closure of $C \cap U^\circ$, choose a net $\langle f_\alpha : \alpha \in D \rangle$ for which each $f_\alpha \in C \cap U^\circ$, and $\lim f_\alpha = f$. Then $f_\alpha \to f$ uniformly on compact sets by Proposition 5.22, so if K is compact, then there exists $\alpha \in D$ s.t. $\beta \succ \alpha \Rightarrow |f_\beta(x) - f(x)| \leq 1$ for all $x \in K$. That is, $f_\beta - f \in K^\circ$. That is, letting K vary, $f_\alpha \to f$ in the τ_K-topology, so $f \in C$. But also $f \in U^\circ$ since U° is weak-* closed. Letting the nets vary, $C \cap U^\circ$ is weak-* closed. Letting U vary, C is almost weak-* closed.

$\square$

The weird countability will come from a recursive application of the following.

Lemma 6.5. *Suppose X is a locally convex space, and W^* is an almost weak-* open neighborhood of 0 in X^*. Suppose U and V are neighborhoods of 0 in X for which $U \subset V$ and $V^\circ \subset W^*$. Then there exists a finite set $F \subset V$ for which $(U \cup F)^\circ \subset W^*$.*

Proof. U° is equicontinuous, as are all $(U \cup F)^\circ$, so $U^\circ - W^*$ and all $(U \cup F)^\circ - W^*$ are relatively weak-* closed in the weak-* compact (Banach-Alaoglu) set U°. So suppose not; suppose all $(U \cup F)^\circ - W^*$ are nonempty. Then since

$$(U \cup F \cup G)^\circ = U^\circ \bigcap F^\circ \bigcap G^\circ$$

$$= (U^\circ \bigcap F^\circ) \bigcap (U^\circ \bigcap G^\circ) = (U \cup F)^\circ \bigcap (U \cup G)^\circ,$$

the family of sets $(U \cup F)^\circ - W^*$ has the finite intersection property. Since $U^\circ - W^*$ is weak-* compact, there exists f which belongs to every $(U \cup F)^\circ - W^*$. Since $f \notin W^* \supset V^\circ$, there exists $x \in V$ such that $|f(x)| > 1$. But this means that $f \notin (U \cup \{x\})^\circ - W^*$.

$\square$

Theorem 6.6 (Banach–Dieudonné). *Suppose X is a first countable Hausdorff locally convex space. Then $\tau_a = \tau_N = \tau_K$.*

Proof. We already know that $\tau_N \subset \tau_K \subset \tau_a$ (Lemma 6.4), and all are translation invariant, so it suffices to show that if W^* is an open τ_a-neighborhood of 0, then W^* contains a τ_N-neighborhood of 0.

Let $U_1 \supset U_2 \supset U_3 \supset \cdots$ be a neighborhood base at 0 for X, and set $U_0 = X$ so that $(U_0)^\circ = \{0\} \subset W^*$. For each n, choose a finite set $F_n \subset U_n$ so that $(U_{n+1} \cup F_0 \cup \cdots \cup F_n)^\circ \subset W^*$; this is done recursively as follows:

Assuming $(U_n \cup (F_0 \cup \cdots F_{n-1})^\circ \subset W^*$, and $U_n \cup (F_0 \cup \cdots \cup F_{n-1}) \supset U_{n+1} \cup (F_0 \cup \cdots \cup F_{n-1})$: there exists $F_n' \subset U_n \cup (F_0 \cup \cdots \cup F_{n-1})$ such that $(U_{n+1} \cup (F_0 \cup \cdots \cup F_{n-1}) \cup F_n')^\circ \subset W^*$, since $U_n \cup (F_0 \cup \cdots \cup F_{n-1})$ and $U_{n+1} \cup (F_0 \cup \cdots \cup F_{n-1})$ are neighborhoods of 0 (Lemma 6.5).

   Set $F_n = F_n' - (F_0 \cup \cdots \cup F_{n-1}) \subset U_n$, giving $U_{n+1} \cup (F_0 \cup \cdots \cup F_{n-1} \cup F_n') = U_{n+1} \cup (F_1 \cup \cdots \cup F_n)$.

Now note that $S = \bigcup F_n$ is the trace of a sequence converging to 0 (go through F_0, then F_1, then $\cdots$). Then for all n,

$$S \cup U_n = U_n \cup (F_0 \cup \cdots \cup F_{n-1})$$

since $F_k \subset U_k \subset U_n$ for $n \geq k$. Hence

$$(S \cup U_n)^\circ = (U_n \cup (F_0 \cup \cdots \cup F_{n-1}))^\circ \subset W^*.$$

But now

$$S^\circ \bigcap U_n^\circ = (S \cup U_n)^\circ \subset W^* \text{ for all } n$$

$$\Rightarrow S^\circ = \bigcup_{n=0}^{\infty}(S^\circ \bigcap U_n^\circ) \subset W^*$$

since $\bigcup U_n^\circ = X^*$ (see the proof of Theorem 6.1). Since S° is a standard τ_N-neighborhood of 0 in X^*, this completes the proof. $\square$

Corollary 6.7. *Suppose X is a Fréchet space. Then any almost* weak-* *continuous linear functional on X^* is evaluation at a point of X.*

Proof. Since X is complete, the closed convex hull of a compact set is compact (Theorem 4.28), so (taking convex hulls) τ_K can be defined using the polars of all compact convex sets. These sets are weakly compact and convex, and the (finer) topology defined by the polars of weakly compact convex sets is precisely the Mackey topology on X^* associated with having evaluation maps be the dual. Since $\tau_K = \tau_a$ is trapped between two topologies (weak-* and Mackey) on X^* having the same dual, τ_a also has this dual. $\square$

Corollary 6.8 (Krein–Smulian I). *Suppose X is a Fréchet space, and C is a convex subset of X^*. Then C is* weak-* *closed in X^* if, and only if, $C \bigcap U^\circ$ is* weak-* *closed for every barrel neighborhood U of 0 in X.*

Proof. This is just Theorem 3.29 applied to (X^*, τ_a). $\square$

   This Krein–Smulian theorem is typically applied to subspaces, but there are other uses as well.

6.3 Properties of the Dual

The two remaining results depend on some unusual consequences of countability in the dual space. As a preliminary, the following is basically a part of the proof that duals of semireflexive spaces are barreled.

Proposition 6.9. *Suppose X is a Hausdorff locally convex space, and suppose D is a weak-* closed, convex, balanced, absorbent subset of X^*. Then D is a strong neighborhood of 0 in X^*.*

Proof. Since D is weak-* closed, convex, balanced, and nonempty, $D = (D_\circ)^\circ$. Set $A = D_\circ$; it suffices to show that A is bounded. But $A^\circ = D$ is absorbent, so A is bounded by Corollary 3.31. □

Now for the countability.

Proposition 6.10. *Suppose X is a Hausdorff, first countable, locally convex space, and suppose V_n is a sequence of convex, balanced, strong neighborhoods of 0 in X^*. Then: If $\bigcap V_n$ absorbs all strongly bounded sets, then $\bigcap V_n$ is a strong neighborhood of 0.*

Proof. Set $V = \bigcap V_n$. Let $U_1 \supset U_2 \supset \cdots$ be a base for the topology of X at 0. Then each U_n° is equicontinuous, hence is strongly bounded. Choose $t_n > 0$ so that $t_n U_n^\circ \subset \frac{1}{2} V$, and choose a bounded set $A_n \subset X$ for which $A_n^\circ \subset \frac{1}{2} V_n$. Set

$$\begin{bmatrix} W_n = \mathrm{con}\,(t_1 U_1^\circ \cup t_2 U_2^\circ \cup \cdots \cup t_n U_n^\circ) + A_n^\circ \\ \subset \frac{1}{2} V + \frac{1}{2} V_n \subset \frac{1}{2} V_n + \frac{1}{2} V_n = V_n \end{bmatrix}$$

since all V_n are convex. Then $\bigcap W_n \subset \bigcap V_n = V$, so it suffices to establish the following lemma, which will be needed later:

Lemma 6.11. *Suppose X is a Hausdorff, first countable, locally convex space. Suppose:*

(α) *$U_1 \supset U_2 \supset \cdots$ is a base for the topology of X at 0,*
(β) *$t_1, t_2, \ldots$ is a sequence of positive real numbers,*
(γ) *$A_1, A_2, \ldots$ is a sequence of bounded subsets of X, and*
(δ) *$W_n = \mathrm{con}\,(t_1 U_1^\circ \cup t_2 U_2^\circ \cup \cdots \cup t_n U_n^\circ) + A_n^\circ$.*

Then $\mathrm{con}(t_1 U_1^\circ \cup t_2 U_2^\circ \cup \cdots \cup t_n U_n^\circ)$ is weak-* compact, and $W = \bigcap W_n$ is a strong neighborhood of 0 in X.

Proof. Each $t_j U_j^\circ$ is weak-* compact (Banach-Alaoglu), so $\mathrm{con}(t_1 U_1^\circ \cup \cdots \cup t_n U_n^\circ)$ is weak-* compact by induction on n. (Usual business: The convex hull of $C \cup D$ is compact when C and D are compact and convex by Proposition 2.14, since it is the continuous image of $C \times D \times [0, 1]$.) Hence W_n is weak-* closed by Corollary 1.15. W_n is also convex and balanced, so W is weak-* closed, convex, and balanced. In view of Proposition 6.9, it suffices to show that W is absorbent. But for all n

and m, there exist constants $s_{m,n} > 0$ such that $s_{m,n} U_n^\circ \subset A_m^\circ$ since each U_n° is equicontinuous, hence is strongly bounded. So: for all m, n we have

$$t_n U_n^\circ \subset W_m \text{ if } m \geq n, \text{ and}$$

$$s_{m,n} U_n^\circ \subset A_m^\circ \subset W_m \text{ if } m < n, \text{ so}$$

$$\varepsilon_n U_n^\circ \subset W_m \text{ for all } m, \text{ where}$$

$$\varepsilon_n = \min(t_n, s_{1,n}, s_{2,n}, \ldots, s_{n-1,n})$$

since W_m is balanced. Hence $\varepsilon_n U_n^\circ \subset \bigcap W_m$, that is $\bigcap W_m$ absorbs all U_n°. Hence $\bigcap W_m$ absorbs all points of $\bigcup U_n^\circ = X^*$.   □

Grothendieck [16] classifies spaces having the intersection property specified in Proposition 6.10; *and* which have a countable sequence of bounded sets B_n (the sets U_n° in the proof) which is *fundamental*, in the sense that any bounded set B is contained in some B_n; as **DF-spaces**. He develops their properties systematically. It is worth looking at.

We can now prove:

Theorem 6.12. *The second dual of a Fréchet space is another Fréchet space.*

Proof. Let $U_1 \supset U_2 \supset \cdots$ be a neighborhood base for the topology of X at 0. Then each U_n° is equicontinuous, hence strongly bounded in X^* (Theorem 4.16(a)). Thus each $U_n^{\circ\circ}$ is a strong neighborhood of 0 in X^{**}. If B is strongly bounded in X^*, then B is equicontinuous [Theorem 4.16(b)], so $B \subset U_n^\circ$ for some n, giving $B^\circ \supset U_n^{\circ\circ}$. That is, the sets $U_n^{\circ\circ}$ form a base for the strong topology of X^{**}, so X^{**} is first countable. It remains to show that X^{**} is sequentially complete. That is where Proposition 6.10 comes in.

Suppose $\langle \Phi_n \rangle$ is a Cauchy sequence in X^{**}. Set $V_n = \{\Phi_n\}_\circ$. Then each V_n is a convex, balanced, strong neighborhood of 0. Also,

$$\bigcap V_n = \bigcap \{\Phi_n\}_\circ = \{\Phi_n : n \in \mathbb{N}\}_\circ$$

absorbs any strongly bounded set D since D° (as a neighborhood of 0 in X^{**}) absorbs $\{\Phi_n : n \in \mathbb{N}\}$: Cauchy sequences are bounded. Hence $V = \bigcap V_n$ is a strong neighborhood of 0 in X^*, and $|\Phi_n(x)| \leq 1$ for all n and all $x \in V$. Since $\Phi(x) = \lim \Phi_n(x)$ exists pointwise as a bounded linear map (Theorem 4.20), this shows that $|\Phi(x)| \leq 1$ for $x \in V$, so Φ is continuous, and $\lim \Phi_n = \Phi$ there (Theorem 4.20 again).   □

There is one thing left.

Theorem 6.13. *Suppose X is a Fréchet space. Then the following are equivalent:*

(i) X^* is barreled.
(ii) X^* is infrabarreled.
iii) X^* is bornological.

Proof. As always, (i) $\Rightarrow$ (ii) $\Leftarrow$ (iii). Also, (ii) $\Rightarrow$ (i) since X^* is complete, by Corollaries 4.22 and 4.8. It remains to prove that (ii) $\Rightarrow$ (iii) for duals of Fréchet spaces. Suppose X^* is infrabarreled, and suppose C is a convex, balanced set that absorbs all bounded sets. The main step is to replace C with a subset that is somewhat less arbitrary.

As before, let $U_1 \supset U_2 \supset \cdots$ be a neighborhood base for X at 0. Each U_n° is equicontinuous, hence is strongly bounded. Choose $t_n > 0$ so that $t_n U_n^\circ \subset C$. Set

$$D_n = \mathrm{con}(t_1 U_1^\circ \cup \cdots \cup t_n U_n^\circ).$$

This set is weak-* compact by Lemma 6.11.

Also, each D_n is convex and balanced, and $D_n \subset C$ since C is convex. Set $D = \bigcup D_n$. Then $D \subset C$. Also D absorbs each U_n°, since $t_n U_n^\circ \subset D_n \subset D$. If B is strongly bounded in X^*, then B is equicontinuous (Theorem 4.16(b) again), so $B \subset U_n^\circ$ for some n, and D absorbs B since it absorbs U_n°. The final claim is that $\frac{1}{2}D^- \subset D$. This will do it, since D, as an ascending union of convex balanced sets, is convex and balanced, so that $\frac{1}{2}D^-$ is closed, convex, balanced, and absorbs all bounded sets, making it a neighborhood of 0 when X^* is infrabarreled. (*Note:* We will then get $\frac{1}{2}D^- \subset D \subset C$, making C a neighborhood of 0.)

The idea for showing that $\frac{1}{2}D^- \subset D$ is to show that if $f \notin D$, then $f \notin \frac{1}{2}D^-$. Suppose $f \notin D$. Then for all n, $f \notin D_n$. Since D_n is weak-* compact (Lemma 6.11), it is weak-* closed, hence is strongly closed. Choose a bounded set $A_n \subset X$ so that $f \notin D_n + A_n^\circ$ in accordance with Proposition 1.9. Set

$$W_n = \frac{1}{2}D_n + A_n^\circ = \mathrm{con}\left(\frac{1}{2}t_1 U_1^\circ \cup \cdots \cup \frac{1}{2}t_n U_n^\circ\right) + A_n^\circ.$$

Then $W = \bigcap W_n$ is a strong neighborhood of 0 by Lemma 6.11. But for all n, since D_n is convex:

$$f \notin D_n + A_n^\circ = \frac{1}{2}D_n + \frac{1}{2}D_n + A_n^\circ = \frac{1}{2}D_n + W_n \supset \frac{1}{2}D_n + W$$

so $f \notin \bigcup\left(\frac{1}{2}D_n + W\right) = \frac{1}{2}D + W$. Hence $f \notin \frac{1}{2}D^-$ by Proposition 1.9. $\square$

Is there an underlying theme for how countability arises in this chapter? The most common characteristic is the ability to form recursive definitions, as has already been noted in Sect. 1.6. Sometimes it is hidden, as in the definition of D_n in the proof of Theorem 6.13. That is not the whole story, though. Recursion does not play any role in the proof of Theorem 6.1. That depends on the Baire category theorem.

But the proof of the Baire category theorem uses a recursive construction . . . .

Hmmm.

Exercises

1. Examine the proof of Theorem 6.1, and verify that (i)–(iii) in Corollary 6.2 are equivalent to:

   (iv) $T^*(X^*)$ is sequentially closed in the strong topology.

2. Suppose X is a Fréchet space, and suppose V is a convex, balanced, absorbent subset of X^* such that $V \cap U^\circ$ is weak-* closed for every barrel U in X. Show that V is a strong neighborhood of 0.

3. (A variant of Theorem 6.1, based on later material.) Suppose X is a Fréchet space, and suppose Z is a strongly closed subspace of X^* with the following property:

$$Z = \bigcup_{n=1}^{\infty} K_n;\ \text{where each } K_n \text{ is weak-* compact, convex, and balanced, and}$$
$$2K_n \subset K_{n+1}.$$

   (a) Show that Z is weak-* closed in X^*.
   (b) Show that, in the proof of Theorem 6.1, this can be arranged for $Z = T^*(Y^*)$ by choosing the base $V_1 \supset V_2 \supset \cdots$ so that each V_n is convex and balanced, and $2V_{n+1} \subset V_n$ (Chap. 3, Exercise 24).

   Note: For part (a), either Krein–Smulian I or Exercise 21 of Chap. 5 is used. This approach to Theorem 6.1 is commonly used.

4. Check that under the hypotheses of Proposition 5.22, $T_\alpha \to T$ uniformly on precompact sets. Then show that under the hypotheses of Theorem 6.6, $\tau_a =$ the topology of uniform convergence on precompact sets. Finally, if X is a Fréchet space, show that τ_a is the Arens topology of Chap. 5, Exercise 23.

5. Suppose X is a Fréchet space, and suppose X^* is separable. Show that X^* is infrabarreled.
   Hint: If D is a countable dense set in X^*, and B is a barrel in X^* that absorbs all strongly bounded sets, then for all $f \in D - B$: $f \notin B = (B^\circ)_\circ$, so choose $\Phi_f \in B^\circ$ so that $|\Phi_f(f)| > 1$. Intersect the sets $\{\Phi_f\}_\circ$ over $f \in D - B$ to form a set W. Show that W is a strong neighborhood of 0 for which $\text{int}(W) \subset B$.

6. (Proposition 3.46 revisited) Show that an LB-space is a DF-space, and show that the strong dual of a DF-space is a Fréchet space. (The latter is a reworking of Theorem 6.12.)

7. One last problem that does not *quite* belong anywhere—except that this chapter can help. Suppose X is a first countable, Hausdorff, locally convex space. If X is a Fréchet space, then the closed convex hull of a compact set is compact; this follows from Theorem 4.28(g), (d), and (h). The purpose of this problem is to show a converse: If the closed convex hull of $\{x_n\} \cup \{0\}$ is weakly compact whenever $x_n \to 0$, then X is complete. To do this, start as usual with a neighborhood base for X at 0 consisting of barrels: $U_1 \supset U_2 \supset U_3 \supset \cdots$. Now set $V_n = 2^{-n}U_n$, so that $2V_{n+1} \subset V_n$ for all n. It suffices, by Theorem 1.35, to show that if $x_n \in V_n$, then Σx_n converges. Suppose $x_n \in V_n$. Then $2^n x_n \in U_n$,

so $\lim 2^n x_n = 0$. Let K be the closed convex hull of $\{2^n x_n\} \cup \{0\}$, and assume K is weakly compact. Set $s_n = \sum_{k=1}^{n} x_k$. Show the following:

(a) $s_n \in K$ for all n.

(b) If $m > n$, then $s_m \in s_n + U_n$.

(c) If s is a weak cluster point of $\langle s_n \rangle$, then $s \in s_n + U_n$ for all n. (Remember: U_n is weakly closed.)

(d) If s is a weak cluster point of $\langle s_n \rangle$, then $s_n \to s$. Since s exists (Proposition 1.5), Σx_n converges. (See Exercise 19 of Chap. 3.)

Now for an approach more in keeping with this chapter. Assume that X is Hausdorff and first countable, and the closed convex hull of $\{x_n\} \cup \{0\}$ is weakly compact whenever $\lim x_n = 0$. With *no* preparation, show the following:

(e) Every τ_N-continuous linear functional on X^* is evaluation at a point of X. (Use Proposition 3.27.)

(f) X is B-complete, hence is complete. (*Hint:* $\tau_N = \tau_a$.)

Notice the difference?

Appendix A
Topological Oddities

This appendix is concerned with those topological results that appear in the text at isolated points and that are not necessarily covered in a beginning graduate course in real analysis. The biggest problem here is organization, since the relationship between subjects is often rather tenuous (and, in one case, nonexistent). Placing subjects in the order they appear in the text would make this discussion almost unreadable, but things will be organized so that deviation from that order is minimal.

The first topic is the uniform convergence of a net of continuous functions. Suppose X is a topological space, (Y, d) is a metric space, and $\langle f_\alpha : \alpha \in D \rangle$ is a net of continuous functions from X to Y. The net $\langle f_\alpha \rangle$ **converges uniformly** to $f : X \to Y$ when the following happens: For each $\varepsilon > 0$, there exists $\alpha \in D$ such that $\beta \succ \alpha$ implies $d(f_\beta(x), f(x)) < \varepsilon$ for all $x \in X$.
Notation: $B(y, r) = $ open ball of radius r around y.

Proposition A.1. *Suppose X is a topological space and (Y, d) is a metric space. Then the uniform limit of a net of continuous functions from X to Y is continuous.*

Proof. As above, suppose $\langle f_\alpha : \alpha \in D \rangle$ is a net of continuous functions from X to Y, which converges uniformly to f. Suppose V is open in Y, and suppose $x_0 \in X$ and $f(x_0) \in V$. We must establish that x_0 is an interior point of $f^{-1}(V)$.

Choose $r > 0$ so that $d(y, f(x_0)) < r \Rightarrow y \in V$. Choose $\alpha \in D$ so that $\beta \succ \alpha \Rightarrow d(f_\beta(x), f(x)) < r/2$ for all $x \in X$. Set $U = f_\alpha^{-1}(B(f(x_0), r/2))$. Then: $\alpha \succ \alpha$, so:

1. $d(f_\alpha(x_0), f(x_0)) < r/2$, so $f_\alpha(x_0) \in B(f(x_0), r/2)$. Hence $x_0 \in U$.
2. If $x \in U$, then $d(f_\alpha(x), f(x_0)) < r/2$, so that

$$d(f(x), f(x_0)) \leq d(f(x), f_\alpha(x)) + d(f_\alpha(x), f(x_0))$$
$$< r/2 + r/2 = r,$$

so that $f(x) \in B(f(x_0), r) \subset V$, and $x \in f^{-1}(V)$.

Put together, this shows that U is an open neighborhood of x_0 in $f^{-1}(V)$. $\qquad \square$

M.S. Osborne, *Locally Convex Spaces*, Graduate Texts in Mathematics 269,
DOI 10.1007/978-3-319-02045-7, © Springer International Publishing Switzerland 2014

The proof above may be familiar; the sequence version is used, for example, in the proof of Urysohn's lemma. The surprise is the appearance of $r/2$ rather than $r/3$, the latter being familiar from the case of functions from $\mathbb{R}$ to $\mathbb{R}$. The reason $r/2$ suffices is that we are not selecting δ's for f_α itself; instead, the openness of $f_\alpha^{-1}(B(f(x_0), r/2))$ is appealed to directly.

Continuing with nets, consider nets in product spaces. Suppose $\langle X_i : i \in \mathscr{I} \rangle$ is a family of topological spaces. An element $\mathbf{x} = \langle x_i \rangle$ in $\prod X_i$ will be denoted with subscripts. To avoid confusion, a net in $\prod X_i$ will be denoted with parentheses.

Proposition A.2. *Suppose* $\langle X_i : i \in \mathscr{I} \rangle$ *is a family of topological spaces, and suppose* $\langle \mathbf{x}(\alpha) : \alpha \in D \rangle$ *is a net in* $\prod X_i$. *Then* $\lim_\alpha \mathbf{x}(\alpha) = \mathbf{x}$ *in* $\prod X_i$ *if and only if* $\lim_\alpha x_i(\alpha) = x_i$ *for all* i.

Proof. Let π_j denote the projection from $\prod X_i$ to X_j. Then $\lim_\alpha \mathbf{x}(\alpha) = \mathbf{x} \Rightarrow$ $\lim_\alpha \pi_j(\mathbf{x}(\alpha)) = \pi_j(\mathbf{x})$ since π_j is continuous [Proposition 1.3(c)], so that $\lim_\alpha x_j(\alpha) = x_j$ for all $j \in \mathscr{I}$.

Suppose $\lim_\alpha x_i(\alpha) = x_i$ for all $i \in \mathscr{I}$. If U is open in $\prod X_i$ and $\mathbf{x} \in U$, then there exists $j_1, \ldots, j_n \in \mathscr{I}$ and open sets $V_1, \ldots, V_n$ ($V_k \subset X_{j_k}$) such that

$$\hat{V} = \prod_{i \in \mathscr{I}} \left\{ \begin{array}{l} V_k \text{ if } i = j_k \text{ for some } k \\ X_i \text{ if not} \end{array} \right\} \subset U$$

and each $x_{j_k} \in V_k$. Choose $\alpha_1, \ldots, \alpha_n \in D$ such that $\beta \succ \alpha_k \Rightarrow x_{j_k}(\beta) \in V_k$, and choose $\alpha \succ D$ such that $\alpha \succ$ every α_k. Then $\beta \succ \alpha \Rightarrow \beta \succ \alpha_k$ for $r = 1, \ldots, n$ so that $\mathbf{x}(\alpha) \in \hat{V} \subset U$.   □

Continuing with products, the next subject is countable products of metric spaces. To start the discussion, observe that any metric space (X, d) has an equivalent, bounded metric

$$d'(x, y) = \frac{d(x, y)}{1 + d(x, y)}.$$

The triangle inequality for d' follows from the discussion preceding Theorem 3.35. This metric is actually bounded by 1, and it not only gives the same topology as d, it also has the same Cauchy sequences as well, since

$$d(x, y) < \varepsilon \Leftrightarrow d'(x, y) < \frac{\varepsilon}{1 + \varepsilon}.$$

The immediate objective is to show that a countable product of metric spaces is metrizable. This can be done directly, but the following generality will be useful later.

Proposition A.3. *Suppose* (X, τ) *is a topological space, and suppose* $\mathscr{C}$ *is a subbase for* τ. *Suppose* $d : X \times X \rightarrow \mathbb{R}$ *is a metric on* X. *Then the* d-*topology coincides with* τ *provided the following two conditions hold:*

(i) For all $x \in X$, $d(x, ?) : X \rightarrow \mathbb{R}$ *is* τ-*continuous.*
(ii) Each $U \in \mathscr{C}$ *is* d-*open.*

Proof. Let τ_d denote the topology produced by d. Then $\mathscr{C} \subset \tau_d$ by (ii), so $\tau \subset \tau_d$. On the other hand, if $x \in X$ and $r > 0$, then each $B(x, r) = d(x, ?)^{-1}((-\infty, r)) \in \tau$ by (i), so $\tau_d \subset \tau$ since the set of all $B(x, r)$'s form a base for τ_d.   □

Corollary A.4. *The countable product of metric spaces is metrizable.*

Proof. Let $\langle (X_n, d_n) : n \in \mathbb{N} \rangle$ be a sequence of metric spaces, with the metrics d_n chosen so that each $d_n \leq 1$. Set

$$d(\langle x_n \rangle, \langle y_n \rangle) = \sum_{n=0}^{\infty} 2^{-n} d_n(x_n, y_n).$$

d is a metric since each d_n is a metric. If $\langle x_n \rangle$ is fixed, then $\langle y_n \rangle \mapsto y_m \mapsto d_m(x_m, y_m)$ is continuous in the product topology, and the series defining d is uniformly convergent (Weierstrass M-test–still valid here), so condition (i) in Proposition A.3 is verified. It remains to check condition (ii).

The subbase to use for (ii) is the usual one for the product topology. Fix m, and suppose U_m is open in X_m. Set

$$\hat{U}_m = \prod_{n=0}^{\infty} \left\{ \begin{array}{l} U_m \text{ if } n = m \\ X_n \text{ if } n \neq m \end{array} \right\}.$$

It remains to show that $\hat{U}_m$ is d-open.

Suppose $\langle x_n \rangle \in \hat{U}_m$, that is $x_m \in U_m$. Choose $r > 0$ so that $B(x_m, r) \subset U_m$. Then

$$2^{-m} r > d(\langle x_n \rangle, \langle y_n \rangle) = \sum_{n=0}^{\infty} 2^{-n} d_n(x_n, y_n)$$

$$\geq 2^{-m} d_m(x_m, y_m)$$

$$\Rightarrow r > d(x_m, y_m) \Rightarrow y_m \in U_m \Rightarrow \langle y_n \rangle \in \hat{U}_m$$

so that $B(\langle x_n \rangle, 2^{-m} r) \subset \hat{U}_m$.   □

Corollary A.5 (Urysohn Metrization Theorem). *Suppose X is a second count-able T_4 space. Then X is metrizable.*

Proof. Let $\mathscr{C}$ be a countable base for the topology of X. Set

$$\mathscr{C}' = \{(B, B') \in \mathscr{C} \times \mathscr{C} : B^- \subset B'\}.$$

$\mathscr{C}'$ is countable; write $\mathscr{C}'$ as $\{(B_1, B_1'), (B_2, B_2'), \ldots\}$. Using Urysohn's lemma, for each n, choose a continuous $f_n : X \to [0, 1]$ such that

$$\forall x \in B_n^- : f_n(x) = 0, \text{ and}$$

$$\forall x \in X - B_n' : f_n(x) = 1.$$

(B_n^- and $X - B_n'$ are two disjoint closed sets.) Note that if $x \neq y$, then $x \in X - \{y\}$, so there exists $B' \in \mathscr{C}$ such that $x \in B' \subset X - \{y\}$. Since $\{x\}$ and $X - B'$ are two disjoint closed sets, there exists disjoint open sets U and V with $x \in U$ and $X - B' \subset V$. Finally, there exists $B \in \mathscr{C}$ with $x \in B \subset U$, so that $B^- \subset U^- \subset X - V \subset B'$. Thus $(B, B') \in \mathscr{C}'$, so that $(B, B') = (B_n, B_n')$ for some n, and $f_n(x) = 0$ while $f_n(y) = 1$.

If X is empty or consists of one point, then X is trivially metrizable, so assume X consists of more than one point. Then the preceding shows that $\mathscr{C}' \neq \emptyset$. Set

$$d(x, y) = \sum_{n=1}^{\infty} 2^{-n} |f_n(x) - f_n(y)|.$$

This function is easily checked to be a metric; the preceding paragraph shows that $x \neq y \Rightarrow d(x, y) > 0$, and the triangle inequality is direct:

$$d(x, z) = \sum_{n=1}^{\infty} 2^{-n} |f_n(x) - f_n(z)|$$

$$\leq \sum_{n=1}^{\infty} 2^{-n} (|f_n(x) - f_n(y)| + |f_n(y) - f_n(z)|)$$

$$= d(x, y) + d(y, z).$$

Furthermore, as before, the series defining $d(x, ?)$ is a uniformly convergent series of continuous functions, so to complete the proof it suffices to show that each $B' \in \mathscr{C}$ is d-open (Proposition A.3). But as before, if $x \in B' \in \mathscr{C}$, then $\{x\}$ and $X - B'$ are disjoint closed sets ... there exists $B \in \mathscr{C}$ with $x \in B \subset B^- \subset B'$, so that $(B, B') = (B_m, B_m')$. Hence

$$2^{-m} > d(x, y) = \sum_{n=1}^{\infty} 2^{-n} |f_n(x) - f_n(y)|$$

$$\geq 2^{-m} |f_m(x) - f_m(y)| = 2^{-m} f_m(y)$$

$$\Rightarrow f_m(y) \neq 1 \Rightarrow y \notin X - B_m' \Rightarrow y \in B_m',$$

so that $B(x, 2^{-m}) \subset B'$.   □

Urysohn is best known for Urysohn's lemma and Urysohn functions. The Urysohn metrization theorem is probably his second most famous result. Less known, but very important, is an obscure fact concerning convergence in a topology, our next subject. Before leaving the metrization theorem, however, it should be remarked that about half the basic texts that prove the Urysohn metrization theorem only assume that X is a second countable T_3 space. However, a second countable T_3 space *is* a T_4 space; see Munkres [26, Theorem 32.1] or Kelley [20, page 113].

Urysohn's convergence result (Urysohn [37]) is not well known, but some similar results are often used in connection with normal families in complex analysis.

Vitali's theorem (Boas [4, p. 217]) provides an example of this. Urysohn's result (generalized slightly) is easy to state and prove, but is rather unintuitive.

Proposition A.6. *Suppose X is a topological space, $\langle x_n \rangle$ is a sequence in X, and $x \in X$. Then $\lim x_n = x$ provided either of the following conditions hold:*

(i) x is a cluster point of every subsequence $\langle x_{n_k} \rangle$ of $\langle x_n \rangle$.
(ii) Every subsequence $\langle x_{n_k} \rangle$ has a further subsequence $\langle x_{n_{k_j}} \rangle$ which converges to x.

Proof. (ii) $\Rightarrow$ (i), since the limit of a subsequence of $\langle x_{n_k} \rangle$ is a cluster point of $\langle x_{n_k} \rangle$. The proof that (i) $\Rightarrow \lim x_n = x$ is by contrapositive;

> *Claim.* If "$\lim x_n = x$" is false, then $\langle x_n \rangle$ has a subsequence $\langle x_{n_k} \rangle$ such that x is not a cluster point of $\langle x_{n_k} \rangle$.

This is straightforward. Assuming "$\lim x_n = x$" is false, there exists an open set U, with $x \in U$, such that x_n is *not* eventually in U. Choose n_1 such that $x_{n_1} \notin U$. Choose $n_2 > n_1$ such that $x_{n_2} \notin U$, and so on. The subsequence $\langle x_{n_k} \rangle$ does not have x as a cluster point since it never enters U. □

For the applications in this book, the fact that (i) implies convergence is most useful. For normal families, it is (ii) that one most often appeals to, using the following:

Corollary A.7. *Suppose X is a sequentially compact Hausdorff space, and suppose $\langle x_n \rangle$ is a sequence in X and $x \in X$. If each convergent subsequence of $\langle x_n \rangle$ converges to x, then $\lim x_n = x$.*

Proof. Each subsequence $\langle x_{n_k} \rangle$ has a further subsequence $\langle x_{n_{k_j}} \rangle$ that converges to some $y \in X$ since X is sequentially compact, and $y = x$ since $\langle x_{n_{k_j}} \rangle$ is a subsequence of $\langle x_n \rangle$. This verifies (ii) in Proposition A.6. □

In Vitali's theorem, for example, the uniqueness of the possible limits of subsequences comes from the identity principle.

Another historical application of Proposition A.6 is the following:

Corollary A.8. *On $[0, 1)$, pointwise convergence a.e. does not coincide with convergence in any topology.*

Proof. Set

$$f_1 = \chi_{[0,1)},$$
$$f_2 = \chi_{[0,\frac{1}{2})}, \qquad f_3 = \chi_{[\frac{1}{2},1)}$$
$$f_4 = \chi_{[0,\frac{1}{3})}, \qquad f_5 = \chi_{[\frac{1}{3},\frac{2}{3})}, \qquad f_6 = \chi_{[\frac{2}{3},1)}$$
$$\cdots$$

$\|f_n\| \to 0$, so each subsequence $\langle f_{n_k} \rangle$ converges to 0 in L^1, hence has a further subsequence $f_{n_{k_j}} \to 0$ a.e. But $f_n \not\to 0$ a.e. □

Continuing the discussion of compactness, the next result is fairly well known, but what does not seem to be well known is that it is actually useful. It does play a role in some proofs of the implicit function theorem.

Proposition A.9 (Compact Graph Theorem). *Suppose X is a compact Hausdorff space, and Y is a Hausdorff space. Suppose $f : X \to Y$ is a function. Then f is continuous if, and only if, its graph $\Gamma(f)$ is compact in $X \times Y$.*

Proof. If f is continuous, then $\Gamma(f) = F(X)$, where $F(x) = (x, f(x)) \in X \times Y$. Since F is continuous and X is compact, $F(X) = \Gamma(f)$ is compact.

Suppose $\Gamma(f)$ is compact. Let $\pi : X \times Y \to X$ be the natural projection. If C is closed in Y, then $X \times C$ is closed in $X \times Y$, so $(X \times C) \bigcap \Gamma(f)$ is compact. Hence $f^{-1}(C) = \pi((X \times C) \bigcap \Gamma(f))$ is compact, hence closed, in X. $\qquad\square$

Corollary A.10. *Suppose X is a compact Hausdorff space, and Y and Z are Hausdorff spaces. Suppose $f : X \to Y$ is continuous and onto, and $g : Y \to Z$ is a function for which $g \circ f$ is continuous. Then g is continuous.*

Proof. $Y = f(X)$ is compact Hausdorff, and for all $y \in Y$, there exists $x \in X$ such that $f(x) = y$, and $g(y) = g \circ f(x)$. This simply says that $\Gamma(g) = F(X)$, where $F(x) = (f(x), g \circ f(x)) \in Y \times Z$. $\qquad\square$

This result will be needed in Appendix C.

Continuing with compactness, the next result is a preliminary to Alexander's lemma. It is also used directly in Theorem 4.28. There is a close analogy between what happens here and some results in commutative ring theory. Consider the following two results:

> **Proposition A.11 (Cohen).** *Suppose R is a commutative ring with identity, and suppose P is an ideal in R which is maximal (under set inclusion) with respect to the property of not being finitely generated. Then P is a prime ideal, and any nonfinitely generated ideal is contained in such a P.*

> **Proposition A.12 (Isaacs).** *Suppose R is a commutative ring with identity, and suppose P is an ideal in R which is maximal (under set inclusion) with respect to the property of not being principal. Then P is a prime ideal, and any nonprincipal ideal is contained in such a P.*

There are others like these. The next result follows the same pattern, provided we make a couple of (nonstandard) definitions.

Suppose $(X, \mathcal{T})$ is a topological space, and suppose $\mathcal{P} \subset \mathcal{T}$. $\mathcal{P}$ will be called an "ideal" if $\mathcal{P}$ is closed under finite unions and swallows intersections from $\mathcal{T}$, that is:

$$U, V \in \mathcal{P} \Rightarrow U \cup V \in \mathcal{P}, \text{ and}$$

$$U \in \mathcal{P}, V \in \mathcal{T} \Rightarrow U \cap V \in \mathcal{P}.$$

$\mathcal{P}$ will be called a "prime ideal" if $U, V \in \mathcal{T}$ and $U \bigcap V \in \mathcal{P}$ implies that $U \in \mathcal{P}$ or $V \in \mathcal{P}$.

Proposition A.13. *Suppose $(X, \mathcal{T})$ is a topological space, and suppose $\mathcal{P}$ is an open cover of X that is maximal (under set inclusion) with respect to the property of not having a finite subcover. Then $\mathcal{P}$ is a prime ideal, and any open cover of X that has no finite subcover is contained in such a $\mathcal{P}$.*

Proof. First, suppose $\mathcal{P}$ is maximal. If $W \in \mathcal{T}$ and $W \notin \mathcal{P}$, then $\mathcal{P} \subsetneqq \mathcal{P} \cup \{W\} \Rightarrow \mathcal{P} \cup \{W\}$ has a finite subcover: $\{W, U_1, \ldots, U_n\}$. Now:

$$U, V \in \mathcal{P} \text{ but } U \cup V \notin \mathcal{P} \Rightarrow \exists \{U \cup V, U_1, \ldots, U_n\}$$
$$\text{(finite subcover)} \Rightarrow \{U, V, U_1, \ldots, U_n\} \text{ is a}$$
$$\text{finite subcover of } \mathcal{P}, \text{ a contradiction,}$$
$$\text{so } U \cup V \in \mathcal{P}.$$

Also,

$$U \in \mathcal{P}, V \in \mathcal{T} \text{ but } U \cap V \notin \mathcal{P} \Rightarrow \exists \{U \cap V, U_1, \ldots, U_n\}$$
$$\text{(finite subcover)} \Rightarrow \{U, U_1, \ldots, U_n\} \text{ is a}$$
$$\text{finite subcover of } \mathcal{P}, \text{ a contradiction,}$$
$$\text{so } U \cap V \in \mathcal{P}.$$

Hence $\mathcal{P}$ is an ideal. $\mathcal{P}$ is prime because

$$U, V \in \mathcal{T}, U \cap V \in \mathcal{P}, U \notin \mathcal{P}, \text{ and } V \notin \mathcal{P}$$
$$\Rightarrow \exists \{U, U_1, \ldots, U_n\} \text{ and } \exists \{V, V_1, \ldots, V_m\},$$
$$\text{finite subcovers of } \mathcal{P} \cup \{U\} \text{ and } \mathcal{P} \cup \{V\},$$
$$\text{respectively. But } x \notin U \Rightarrow x \in \text{ some } U_j$$
$$\text{and } x \notin V \Rightarrow x \in \text{ some } V_k, \text{ so } \{U \cap V, U_1, \ldots, U_n, V_1, \ldots, V_m\}$$
$$\text{is a finite subcover of } \mathcal{P}, \text{ a contradiction.}$$

Now suppose $\mathcal{C}$ is an open cover of X that has no finite subcover. Set

$$\mathcal{A} = \{\mathcal{C}' : \mathcal{C} \subset \mathcal{C}' \subset \mathcal{T} \text{ and } \mathcal{C}' \text{ has no finite subcover of } X.\}.$$

$\mathcal{C} \in \mathcal{A}$, so $\mathcal{A} \neq \emptyset$. If $\mathcal{D}$ is a nonempty chain in $\mathcal{A}$, set $\mathcal{C}_0 = \bigcup \mathcal{D}$. Then $\mathcal{C} \subset \mathcal{C}_0 \subset \mathcal{T}$. If $\mathcal{C}_0$ has a finite subcover $\{U_1, \ldots, U_n\}$, then each $U_j \in \mathcal{C}_j \in \mathcal{D}$, for some $\mathcal{C}_j$, so there exists l with each $U_j \in \mathcal{C}_l$ since $\mathcal{D}$ is totally ordered. But now $\{U_1, \ldots, U_n\}$ is a finite subcover of $\mathcal{C}_l$, a contradiction. Hence $\mathcal{C}_0 \in \mathcal{A}$. This verifies the hypotheses of Zorn's lemma, so $\mathcal{A}$ has a maximal element $\mathcal{P}$. $\qquad \square$

A property of prime ideals is the following.

Lemma A.14. *Suppose $(X, \mathcal{T})$ is a topological space, $\mathcal{P}$ is a prime ideal in $\mathcal{T}$, and $\mathcal{C}$ is a subbase for $\mathcal{T}$. Then $\bigcup \mathcal{P} = \bigcup (\mathcal{P} \cap \mathcal{C})$.*

Proof. $\bigcup \mathcal{P} \supset \bigcup(\mathcal{P} \cap \mathcal{C})$ trivially. Suppose $x \in \bigcup \mathcal{P}$, that is there exists $U \in \mathcal{P}$ such that $x \in U$. There exists $V_1, \ldots, V_n \in \mathcal{C}$ such that $x \in V_1 \cap \cdots \cap V_n \subset U$ since $\mathcal{C}$ is a subbase. Now $V_1 \cap \cdots \cap V_n = V_1 \cap \cdots \cap V_n \cap U \in \mathcal{P}$ since $\mathcal{P}$ is an ideal, and there exists j with $V_j \in \mathcal{P}$ since $\mathcal{P}$ is prime. Hence $x \in V_j \in \mathcal{P} \cap \mathcal{C}$. $\qquad \square$

Corollary A.15 (Alexander's Lemma). *Suppose $(X, \mathcal{T})$ is a topological space, and $\mathcal{C}$ is a subbase for $\mathcal{T}$. If every open cover from $\mathcal{C}$ has a finite subcover, then X is compact.*

Proof. Suppose $(X, \mathcal{T})$ is not compact. Then there exists a $\mathcal{P}$ as in Proposition A.13, so that $\mathcal{P} \cap \mathcal{C}$ will be a cover of X (Lemma A.14) with no finite subcover.   □

Theorem A.16 (Tychenoff Product Theorem). *The product of any family of compact topological spaces is compact.*

Proof. Suppose $\langle X_i : i \in \mathcal{I} \rangle$ is a family of compact spaces. For each $j \in \mathcal{I}$, let $\mathcal{C}_j$ denote the subbase elements of the form

$$\hat{U}_j = \prod_{i \in \mathcal{I}} \left\{ \begin{array}{l} U_j \text{ if } i = j \\ X_i \text{ if } i \neq j \end{array} \right\}$$

manufactured from the open subsets U_j of X_j. Then $\bigcup \mathcal{C}_i$ is a subbase for the product topology, so by Alexander's lemma it suffices to show that an open cover from $\bigcup \mathcal{C}_i$ has a finite subcover. Let $\mathcal{D}$ be an open cover from $\bigcup \mathcal{C}_i$, and set

$$\mathcal{D}_i = \mathcal{D} \cap \mathcal{C}_i; \mathcal{D}_i' = \{U_i \subset X_i : \hat{U}_i \in \mathcal{D}_i\}.$$

There are now two possibilities.

1. Some $\mathcal{D}_i'$ covers X_i. But then $X_i \subset U_i(1) \cup \cdots \cup U_i(n) \Rightarrow X \subset \hat{U}_i(1) \cup \cdots \cup \hat{U}_i(n)$, giving a finite subcover.
2. No $\mathcal{D}_i'$ covers X_i. But then for all j one can choose x_j such that $x_j \notin U_j$ for all $U_j \in \mathcal{D}_j'$. Thus, $\langle x_i \rangle \notin \hat{U}_j$ for any $U_j \in \mathcal{D}_j'$ or any j. That means that $\bigcup \mathcal{D}_i = \mathcal{D}$ does not cover X.

                                                                                          □

Now for local compactness. A compact Hausdorff space is locally compact by definition, and an open subset of a locally compact Hausdorff space is locally compact since Proposition 1.6(i) holds. When discussing topological measures on locally compact Hausdorff spaces, there are two routes one can take.

1. Baire measures. The class of Baire sets is the σ-algebra generated by the compact G_δ's. The primary advantage here is that Baire measures are automatically regular on the σ-bounded Baire sets. Royden [30] uses this approach.
2. Radon measures on the Borel sets. A Radon measure is a Borel measure that is inner regular on open sets and outer regular on Borel sets. The primary advantage here is that the theory of supports becomes available. Folland [15] uses this approach.

Given a Baire measure μ, there is a unique Radon measure $\hat{\mu}$ that agrees with μ on the σ-bounded Baire sets (Royden [30, Theorem 22, Chap. 13]), and this is necessary for a satisfactory theory of supports, since the support of a measure need not be a Baire set. The support is defined as follows. Suppose μ is a Radon measure (or even a Baire measure). Let Z denote the class of open sets U (or open Baire sets) such that $\mu(U) = 0$. Z is closed under countable unions. Set

$$\text{supp}(\mu) = X - \bigcup_{U \in Z} U.$$

If K is compact (or just a compact Baire set), and $K \subset X - \text{supp}(\mu)$, then Z covers K, so K is contained in a finite union of members of Z, and so $K \subset U$ for some $U \in Z$. Hence $\mu(K) \leq \mu(U) = 0$. Since μ is inner regular on open sets, $\mu(X - \text{supp}(\mu)) = 0$. [This final conclusion is unavailable if μ is only a Baire measure and $\text{supp}(\mu)$ is not a Baire set.] Furthermore, if V is open and $V \cap \text{supp}(\mu) \neq \emptyset$, then by definition $V \notin Z$, so $\mu(V) > 0$. Since $V - \text{supp}(\mu) \subset X - \text{supp}(\mu)$: $\mu(V - \text{supp}(\mu)) = 0$. This proves the following:

Proposition A.17. *Suppose X is a locally compact Hausdorff space and μ is a Radon measure on X. Suppose U is open in X. Then:*

(a) If $U \cap \text{supp}(\mu) = \emptyset$, then $\mu(U) = 0$.
(b) If $U \cap \text{supp}(\mu) \neq \emptyset$, then $\mu(U) = \mu(U \cap \text{supp}(\mu)) > 0$.

There is a corollary that will be needed in Appendix C.

Corollary A.18. *Suppose X is a compact Hausdorff space, and let $\mathcal{K}$ denote the set of positive Radon measures μ on X for which $\mu(X) = 1$, considered as a (convex) subset of $C(X)^*$ via the Riesz representation theorem. Then the extreme points of $\mathcal{K}$ are the point measures μ_p for $p \in K$:*

$$\mu_p(E) = \begin{cases} 1 \text{ if } p \in E \\ 0 \text{ if } p \notin E \end{cases}.$$

Proof. Each μ_p is extreme: If $\mu_p = t\mu + (1-t)\nu, 0 < t < 1$, with $\mu, \nu \in \mathcal{K}$, then $0 = \mu_p(X - \{p\}) = t\mu(X - \{p\}) + (1-t)\nu(X - \{p\})$, so $\mu(X - \{p\}) = \nu(X - \{p\}) = 0$. Hence $\mu(\{p\}) = \mu(X) = 1 = \nu(X) = \nu(\{p\})$, and $\mu = \nu = \mu_p$.

An extreme μ has the form μ_p for some p: By Proposition A.17(a), $\text{supp}(\mu) = \emptyset$ is out [otherwise, $\mu(X) = 0$], so $\text{supp}(\mu) \neq \emptyset$. If $p \neq q$; $p, q \in \text{supp}(\mu)$, choose disjoint open U and V with $p \in U$ and $q \in V$. Then $\mu(U) > 0$ and $\mu(V) > 0$ by Proposition A.17(b), so

$$\mu = \mu(U) \cdot \nu + (1 - \mu(U)) \cdot \lambda; \text{ where}$$

$$\nu(E) = \mu(E \cap U)/\mu(U), \text{ and}$$

$$\lambda(E) = \mu(E - U)/\mu(X - U).$$

$\nu(U) = 1$ and $\nu(V) = 0$, so ν is not μ, and μ is not extreme. Thus, if μ is extreme, then $\text{supp}(\mu) = \{p\}$ consists of one point. Hence $\mu = \mu_p$ by the considerations in the first part of this proof. $\qquad\square$

While not strictly needed here, the next result illuminates the situation where supports are not Baire sets. It also is directly connected with its corollary, which is relevant to our next subject.

Proposition A.19. *Suppose X is a locally compact Hausdorff space. Then any compact Baire set is a G_δ.*

Proof. Suppose K is a compact Baire set. Then K is contained in the σ-algebra generated by the compact G_δ's, so there exists compact G_δ's $K_1, K_2, \ldots$ such that K is contained in the σ-algebra generated by $\{K_1, K_2, \ldots\}$. For each n, choose a continuous $f_n : X \to [0, 1]$ such that $K_n = f_n^{-1}(\{1\})$. Define

$$\mathbf{f} = \langle f_n \rangle : X \to Y = \prod_{n=1}^{\infty} [0, 1].$$

This $\mathbf{f}$ is continuous since its coordinate functions are continuous. Moreover,

$$K_m = \mathbf{f}^{-1}\left(\prod_{n=1}^{\infty} \left\{ \begin{matrix} \{1\} & \text{if } n = m \\ [0, 1] & \text{if } n \neq m \end{matrix} \right\} \right).$$

Let $\mathscr{B}_0(Y)$ denote the Borel sets in Y; then $\mathbf{f}^{-1}(\mathscr{B}_0(Y))$ is a σ-algebra in X that contains every K_m, so $K \in \mathbf{f}^{-1}(\mathscr{B}_0(Y))$. That is, $K = \mathbf{f}^{-1}(A)$ for some Borel set $A \subset Y$. Hence $\mathbf{f}(K) \subset A$, so that

$$K = \mathbf{f}^{-1}(A) \supset \mathbf{f}^{-1}(\mathbf{f}(K)) \supset K$$

and $K = \mathbf{f}^{-1}(\mathbf{f}(K))$. But $\mathbf{f}(K)$ is compact and Y is metrizable (Corollary A.4), so $\mathbf{f}(K)$ is a G_δ in Y. (Any closed subset B of a metric space is a G_δ: Set $U_n = \{x : d(x, B) < \frac{1}{n}\}$.) Writing $\mathbf{f}(K) = \bigcap U_n$ gives $K = \bigcap \mathbf{f}^{-1}(U_n)$.   □

Corollary A.20. *Suppose X is a locally compact Hausdorff space, and $p \in X$. Then $\{p\}$ is a Baire set if, and only if, there is a countable neighborhood base at p.*

Proof. If there is a countable neighborhood base $\{V_1, V_2, \ldots\}$ at p, then $\{p\} = \bigcap \text{int}(V_j)$ is a G_δ, hence is a Baire set. Suppose $\{p\}$ is a Baire set, so that $\{p\} = \bigcap U_n$ for a sequence of open sets U_n. Using Proposition 1.6, choose for each n a compact neighborhood V_n of p such that $p \in V_n \subset U_n$. Set $W_n = V_1 \cap \cdots \cap V_n$. $p \in \text{int}(V_1) \cap \cdots \cap \text{int}(V_n) \subset W_n$, so each W_n is a neighborhood of p. The claim is that $\{W_n\}$ is actually a neighborhood base at p.

Suppose $p \in U$ and U is open. If no $W_n \subset U$, then for all n, $W_n - U \neq \emptyset$. Since $W_1 - U = V_1 - U$ is compact and each $W_n - U$ is closed, by Cantor intersection

$$\left(\bigcap_{n=1}^{\infty} W_n \right) - U = \bigcap_{n=1}^{\infty} (W_n - U) \neq \emptyset.$$

But this cannot happen, since $W_n \subset V_n \subset U_n$, and $\bigcap U_n = \{p\} \subset U$. Hence there exists $n : W_n \subset U$.   □

Compact Hausdorff spaces that are not first countable are easily constructed; the product of uncountably many copies of $[0, 1]$ is such a space. This leads to an additional class of examples.

Proposition A.21. *Suppose X is a compact Hausdorff space that is not first countable. Let p be a point of X for which there is no countable neighborhood base. Then*

$$X \times [0, 1] - \{(p, 0)\}$$

is a locally compact Hausdorff space that is not normal.

Proof. The two closed sets that cannot be separated are $A = \{p\} \times (0, 1]$ and $B = (X - \{p\}) \times \{0\}$. The reason is that if U is open and $A \subset U$, then $B \cap U^- \neq \emptyset$. Suppose U is open and $A \subset U$. For clarity, replace U with $U \cap (X \times (0, 1])$, which is still open and contains A. For each $x \in (0, 1]$, choose open neighborhoods V_x of p in X and W_x of x in $(0, 1]$ for which $V_x \times W_x \subset U$; this is possible since our new U is open in $X \times (0, 1]$. The sets $\{W_x\}$ cover $(0, 1]$ and $(0, 1]$ is Lindelöf, so there exists $x_1, \ldots, x_n, \ldots$ such that $(0, 1] = \bigcup W_{x_n}$. Set $V = \bigcap V_{x_n}$. Then $\{p\} \subsetneq V$ since $\{p\}$ is not a G_δ (Proposition A.19 and Corollary A.20). But: If $y \in (0, 1]$, then there exists n such that $y \in W_{x_n}$, so

$$V \times \{y\} \subset V_{x_n} \times W_{x_n} \subset U.$$

Hence $V \times (0, 1] \subset U$, and $B \cap U^- \supset B \cap (V \times (0, 1])^- \supset (V - \{p\}) \times \{0\}$.   □

Now for the final topic, which is not really topology, and does not connect with anything. It concerns imbedding a normed space in its completion, as described in the proof of Proposition 5.28. The situation there was that X was a normed space, and the isomorphic (and isometric) copy $J_X(X)$ was a literal subspace of its completion (which was its closure in X^{**}). The construction here is general, and applies when we have a mathematical object X which is isomorphic to a mathematical object Y, which (in turn) imbeds in a mathematical object $\hat{Y}$. The idea is to produce an isomorphic extension $\hat{X}$ of X of the same type:

$$\hat{X} \approx \hat{Y}$$
$$\cup \quad \cup$$
$$X \approx Y$$

Examples of this kind of thing abound; the usual proof that any metric space X has a completion actually produces a completion of an isometric copy of X. Similarly, the proof in algebra that a polynomial over a field has a splitting field actually produces an extension of an isomorphic copy of the field. The same applies to quotient fields of integral domains, and this illustrates where the problem lies.

Start with $\mathbb{Z}$, the integers. To get to $\mathbb{C}$, first construct $\mathbb{Q}$ (the quotient field), then $\mathbb{R}$ (the completion of $\mathbb{Q}$), then $\mathbb{C}$ (the splitting field of $x^2 + 1$). When carrying out the last part, the ordered pair $(1, 2)$ will be $1 + 2i$. Or will it? $(1, 2)$ *is taken*! It *already* was used in the first part to stand for $\frac{1}{2}$. The point is that we cannot just take $\hat{X}$ to be $X \bigcup (\hat{Y} - Y)$ since X and $\hat{Y} - Y$ may already overlap.

Here is how to do it. Construct an object $\hat{Z}$ which is isomorphic to $\hat{Y}$ and which is (set theoretically) disjoint from everything already used $(X \bigcup \hat{Y})$. We now have

$$\tilde{Y} \approx \hat{Z}$$

$$\cup \quad \cup$$

$$X \approx Y \approx Z$$

$$\boxed{\begin{array}{l} \text{Image of} \\ Y \text{ under} \\ \hat{Y} \xrightarrow{\approx} \hat{Z} \end{array}}$$

$$\longleftarrow$$

Shorten this to

$$\hat{Z}$$

$$\cup$$

$$X \approx Z$$

and set $\hat{X} = X \bigcup (\hat{Z} - Z)$. Copy the structure of $\hat{Z}$ to $\hat{X}$; roughly speaking, remove Z from $\hat{Z}$ and replace it with X:

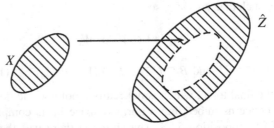

$\hat{Z}$ gets an "object transplant."

Any time the preceding makes sense, this construction can be carried out. While it is possible to be precise about this, the required subject is category theory. However, in most cases, it is fairly clear what is needed:

1. Our mathematical objects are sets with additional structure.
2. Our isomorphisms are bijective set functions.
3. Given a set $\hat{Z}$ in one-to-one correspondence with our object, the entire structure of our object can be copied onto $\hat{Z}$.

Appendix B
Closed Graphs in Topological Groups

It has already been noted in Sect. 4.4 that a version of the open mapping theorem can be proved for topological groups. The same holds for the closed graph theorem. Both follow from a single, rather complicated lemma (Lemma B.2 below). The results here are not very well known, although they do appear in Kelley [20, pp. 213–4]. The problem is that the results are not really as useful for topological groups because there is no analog of the "barreled" condition.

The first result isolates what a closed graph actually does here. (Recall that the graph of a homomorphism is a subgroup of the product.) In the proof of Theorem 4.37, this was the last step. Although the approach here is modeled on Theorem 4.37, it seems best to clarify at the start what "closed graph" means for us.

Proposition B.1. *Suppose G and $\tilde{G}$ are topological groups, and H is a closed subgroup of $G \times \tilde{G}$. Let*

$$\pi : G \times \tilde{G} \to G \text{ and}$$
$$\tilde{\pi} : G \times \tilde{G} \to \tilde{G}$$

denote the canonical projections, and set $\varphi = \pi|H$, $\phi = \tilde{\pi}|H$. Let $\mathscr{B}_e$ denote a neighborhood base at the identity e of G. Then for all $g \in G$:

$$\phi(\varphi^{-1}(g)) = \bigcap_{B \in \mathscr{B}_e} \phi(\varphi^{-1}(gB))^-.$$

Proof. $g \in gB$, so $\varphi^{-1}(g) \subset \varphi^{-1}(gB)$, so $\phi(\varphi^{-1}(g)) \subset \phi(\varphi^{-1}(gB)) \subset \phi(\varphi^{-1}(gB))^-$ for all $B \in \mathscr{B}_e$. Hence $\phi(\varphi^{-1}(g))$ is contained in the intersection.

Suppose $\tilde{g} \in \phi(\varphi^{-1}(gB))^-$ for all $B \in \mathscr{B}_e$. Suppose U is open in G and V is open in $\tilde{G}$, with $g \in U$ and $\tilde{g} \in V$. Then there exists $B \in \mathscr{B}_e$ with $gB \subset U$. Now $\tilde{g} \in \phi(\varphi^{-1}(gB))^-$, so $V \cap \phi(\varphi^{-1}(gB)) \neq \emptyset$, say $y = \phi((x, y)) \in V$ $((x, y) \in \varphi^{-1}(gB) \cap H)$, and $x = \varphi((x, y)) \in gB \subset U$ (since $(x, y) \in \varphi^{-1}(gB)$). Then $(x, y) \in U \times V \cap H$.

We have just demonstrated that if U is open in G and V is open in $\tilde{G}$, with $(g, \tilde{g}) \in U \times V$, then $U \times V \cap H \neq \emptyset$. Thus, $(g, \tilde{g}) \in H$, since H is closed. But now $(g, \tilde{g}) \in \varphi^{-1}(g)$, and $\tilde{g} \in \phi(\varphi^{-1}(g))$.   □

Now for the complicated lemma.

Lemma B.2. *Suppose G, $\tilde{G}$, and H are Hausdorff topological groups, with G being first countable and complete. Suppose $\mathscr{B}_e = \{B_1, B_2, \ldots\}$ is a neighborhood base at the identity e of G, consisting of closed sets, satisfying $B_j = B_j^{-1} \supset B_{j+1}^2$. Suppose $\varphi : H \to G$ and $\psi : H \to \tilde{G}$ are two homomorphisms satisfying the following two conditions:*

$$\forall g \in G : \psi(\varphi^{-1}(g)) = \bigcap_{j=1}^{\infty} \psi(\varphi^{-1}(gB_j))^{-} \qquad (*)$$

$$\forall j : \psi(\varphi^{-1}(B_j))^{-} \text{ is a neighborhood of the identity } \tilde{e} \text{ of } \tilde{G}. \qquad (**)$$

Then: $\psi(\varphi^{-1}(B_1)) \supset \psi(\varphi^{-1}(B_2))^{-}$.

Proof. The idea is this. Suppose $\tilde{g} \in \psi(\varphi^{-1}(B_2))^{-}$. We shall produce elements $x_n \in B_n$, with $x_1 = e$, so that $y = \prod x_n$ converges. Now $x_1 x_2 \cdots x_n = x_2 x_3 \cdots x_n \in B_2 \cdots B_n \subset B_1$, so $y \in B_1$ since B_1 is closed. Finally, we show there exists $h \in H$ with $\varphi(h) = y$ and $\psi(h) = \tilde{g}$ (i.e., $\tilde{g} \in \psi(\varphi^{-1}(y)) \subset \psi(\varphi^{-1}(B_1)))$ by showing that $\tilde{g} \in \psi(\varphi^{-1}(yB_j))^{-}$ for all j, and then applying $(*)$.

To start with,

$$\tilde{g} \in \psi(\varphi^{-1}(B_2))^{-} \subset \psi(\varphi^{-1}(B_2)) \cdot \psi(\varphi^{-1}(B_3))^{-}$$

by Proposition 1.9, since $\psi(\varphi^{-1}(B_3))^{-}$ is a neighborhood of $\tilde{e}$. Hence, we can choose $x_2 \in B_2$, and $h_2 \in H$, for which $\varphi(h_2) = x_2$ and $\tilde{g} \in \psi(h_2)\psi(\varphi^{-1}(B_3))^{-}$, that is

$$\psi(h_2)^{-1}\tilde{g} \in \psi(\varphi^{-1}(B_3))^{-}.$$

Now repeat: Given $x_1 = e$ and $x_2 \cdots x_n$, and $h_2 \cdots h_n$ for which $\varphi(h_j) = x_j$ and

$$\psi(h_n)^{-1} \cdots \psi(h_2)^{-1}\tilde{g} \in \psi(\varphi^{-1}(B_{n+1}))^{-} \subset \psi(\varphi^{-1}(B_{n+1})) \cdot \psi(\varphi^{-1}(B_{n+2}))^{-},$$

choose $x_{n+1} \in B_{n+1}$, and $h_{n+1} \in H$, with $\varphi(h_{n+1}) = x_{n+1}$, and

$$\psi(h_n)^{-1} \cdots \psi(h_2)^{-1}\tilde{g} \in \psi(h_{n+1})\psi(\varphi^{-1}(B_{n+2}))^{-}$$

so that

$$\psi(h_{n+1})^{-1} \cdots \psi(h_2)^{-1}\tilde{g} \in \psi(\varphi^{-1}(B_{n+2}))^{-}.$$

Now form the infinite product $y = \prod x_n$. We need to show that $\tilde{g} \in \psi(\varphi^{-1}(yB_j))^{-}$ for all j.

Fix any positive integer j. Note that for $n \geq j + 3$, $x_{j+2}x_{j+3} \cdots x_n \in B_{j+2}B_{j+3} \cdots B_n \subset B_{j+1}$, so

$$
x_{j+1}^{-1} x_j^{-1} \cdots x_2^{-1} y = \lim_{n \to \infty} x_{j+1}^{-1} x_j^{-1} \cdots x_2^{-1} x_2 \cdots x_j x_{j+1} \cdots x_n
$$
$$
= \lim_{n \to \infty} x_{j+2} \cdots x_n \in B_{j+1}
$$

since B_{j+1} is closed. Hence $y^{-1} x_2 \cdots x_{j+1} = (x_{j+1}^{-1} \cdots x_2^{-1} y)^{-1} \in B_{j+1}^{-1} = B_{j+1}$, so $x_2 \cdots x_{j+1} \in y B_{j+1}$. Now suppose $\varphi(h) \in B_{j+2}$. Then $\varphi(h_2 \cdots h_{j+1} h) = \varphi(h_2) \cdots \varphi(h_{j+1}) \varphi(h) = x_2 \cdots x_{j+1} \varphi(h) \in y B_{j+1} B_{j+2} \subset y B_j$. That is, $h_2 \cdots h_{j+1} h \in \varphi^{-1}(y B_j)$. Since h was arbitrary in $\varphi^{-1}(B_{j+2})$, we get that $h_2 \cdots h_{j+1} \varphi^{-1}(B_{j+2}) \subset \varphi^{-1}(y B_j)$. But by construction; and noting that $\psi(hA) = \psi(h)\psi(A)$ for any $h \in H$ and $A \subset H$, and $(\tilde{x} B)^- = \tilde{x}(B^-)$ for any $\tilde{x} \in \tilde{G}$ and $B \subset \tilde{G}$:

$$
\psi(h_{j+1})^{-1} \cdots \psi(h_2)^{-1} \tilde{g} \in \psi(\varphi^{-1}(B_{j+2}))^-, \text{ so}
$$
$$
\tilde{g} \in \psi(h_2) \cdots \psi(h_{j+1}) \psi(\varphi^{-1}(B_{j+2}))^-
$$
$$
= \psi(h_2 \cdots h_{j+1}) \psi(\varphi^{-1}(B_{j+2}))^-
$$
$$
= (\psi(h_2 \cdots h_{j+1}) \psi(\varphi^{-1}(B_{j+2}))^-
$$
$$
= \psi(h_2 \cdots h_{j+1} \varphi^{-1}(B_{j+2}))^-
$$
$$
\subset \psi(\varphi^{-1}(y B_j))^-
$$

as required.   □

To formulate the open mapping and closed graph theorems for topological groups, we need two definitions, one of which has already appeared in disguise (Corollary 4.33).

Definition B.3. Suppose G and $\tilde{G}$ are Hausdorff topological groups, with identity elements e and $\tilde{e}$, respectively. Suppose $f : G \to \tilde{G}$ is a homomorphism. Then f is called **nearly continuous** if $f^{-1}(\tilde{U})^-$ is a neighborhood of e whenever $\tilde{U}$ is a neighborhood of $\tilde{e}$, and f is called **nearly open** if $f(U)^-$ is a neighborhood of $\tilde{e}$ whenever U is a neighborhood of e.

Theorem B.4. *Suppose G and $\tilde{G}$ are Hausdorff topological groups, and $f : G \to \tilde{G}$ is a homomorphism whose graph is closed.*

(a) *(CGT) Assume G is first countable and complete. If f is nearly continuous, then f is continuous.*

(b) *(OMT) Assume $\tilde{G}$ is first countable and complete. If f is nearly open, then f is open.*

Proof. (b) first. Assume f is nearly open. Let H denote the graph of f, denoted by $H = \Gamma(f)$, and

$$\pi : G \times \tilde{G} \to G$$

$$\tilde{\pi} : G \times \tilde{G} \to \tilde{G}$$

$$\varphi = \pi|_H, \psi = \tilde{\pi}|_H$$

as in Proposition B.1. Now let U be any neighborhood of e. Choose a neighborhood base at e, $\mathscr{B}_e = \{B_1, B_2, \ldots\}$, in accordance with Theorem 1.13, in such a way that $B_j = B_j^{-1} \supset B_{j+1}^2$, all B_j are closed, and $B_1 \subset U$. Then condition $(*)$ holds, thanks to Proposition B.1.

Suppose $A \subset G$. Then

$$\varphi^{-1}(A) = \{(x, f(x)) \in G \times \tilde{G} : x \in A\}$$

$$\text{and } \psi(\varphi^{-1}(A)) = \{f(x) : x \in A\}$$

$$= f(A).$$

That is, condition $(**)$ is simply the statement that $f(B_j)^-$ is a neighborhood of $\tilde{e}$ for all j, which is also true since we are assuming that f is nearly open. We now have that $f(U) \supset f(B_1) \supset f(B_2)^-$, a neighborhood of $\tilde{e}$. Hence f is an open mapping by Proposition 1.26(b).

(a) Assume f is nearly continuous. Here, we have to reverse the roles of G and $\tilde{G}$, so $\tilde{G}$ is first countable and complete, while we work with $\varphi(\psi^{-1}(\text{sets in } \tilde{G}))$. Now let $\tilde{U}$ be any neighborhood of $\tilde{e}$. Choose a neighborhood base at $\tilde{e}$. $\mathscr{B}_{\tilde{e}} = (\tilde{B}_1, \tilde{B}_2, \ldots)$, in accordance with Theorem 1.13 in such a way that $\tilde{B}_j = \tilde{B}_j^{-1} \supset \tilde{B}_{j+1}^2$, all $\tilde{B}_j$ are closed, and $\tilde{B}_1 \subset \tilde{U}$. Then condition $(*)$ holds, thanks to Proposition B.1.

Suppose $\tilde{A} \subset \tilde{G}$. Then

$$\psi^{-1}(\tilde{A}) = \{(x, f(x)) \in G \times \tilde{G} : f(x) \in \tilde{A}\}$$

$$\text{and } \varphi(\psi^{-1}(\tilde{A})) = \{x \in G : f(x) \in \tilde{A}\}$$

$$= f^{-1}(\tilde{A}).$$

That is, condition $(**)$ becomes simply the statement that $f^{-1}(\tilde{B}_j)^-$ is a neighborhood of e for all j, which is also true since we are assuming that f is nearly continuous. We now get that $f^{-1}(\tilde{U}) \supset f^{-1}(\tilde{B}_1) \supset f^{-1}(\tilde{B}_2)^-$, a neighborhood of e. Hence f is continuous by Proposition 1.26(a). $\square$

Of course, verifying all that "nearly" business puts a severe crimp in applications to topological groups. However, it does clarify at least two things:

1. There really is an underlying theme to the closed graph theorem and the open mapping theorem, even in general; namely, Lemma B.2.
2. The "barreled" condition, available for functional analysis, is *really* helpful there.

Left as an exercise is the analog of Corollary 4.33:

A homomorphism $f : G \to \tilde{G}$, where G and $\tilde{G}$ are topological groups, is nearly continuous if, and only if, it has the (purely topological) property that $f^{-1}(\tilde{U}) \subset \mathrm{int}(f^{-1}(\tilde{U})^-)$ for all open $\tilde{U} \subset \tilde{G}$.

Appendix C
The Other Krein–Smulian Theorem

The Krein–Smulian theorem, which appears in Sect. 6.2, is not the only Krein–Smulian theorem. There is another, which states (in its simplest form) that in a quasi-complete Hausdorff locally convex space, the closed convex hull of a weakly compact set is weakly compact. In functional analysis textbooks that deal primarily with Banach spaces, this version is *the* Krein–Smulian theorem, although some books (e.g., Dunford and Schwartz [12] and Conway [7]) cover both versions. In textbooks that treat general spaces, the Krein–Smulian theorem is Corollary 6.8, and the result to be discussed in this appendix is simply called Krein's theorem.

There actually is a reason for this. The result was first proved by Krein [23] for separable Banach spaces and then generalized to all Banach spaces in a joint paper by Krein and Smulian [24]. The generalization to quasi-complete spaces seems to be due to Grothendieck [16], who simply called it Krein's theorem. The name "Krein–Smulian II" is used here simply because it is so generally recognized as a "Krein–Smulian" theorem.

There are basically two approaches, both of which require some version of Eberlein's theorem. One uses iterated limits, and a combined version of Eberlein's theorem with Krein–Smulian II appears as Theorem 17.12 in Kelley and Namioka [21]. The iterated limit gimmick is due to Grothendieck, and the proof is direct but very messy. The approach here is sometimes called the "integral" approach, although the interpretation as an integral is only after the fact; the relevant map is really an adjoint. The primary advantage is that Eberlein's theorem is only needed for Banach spaces, where a much simpler proof is available; the approach below to Eberlein's theorem is a hybrid of the argument by Dunford and Schwartz [12] and that of Narici and Beckenstein [27].

Lemma C.1. *Suppose X is a normed space, and suppose Y is a finite-dimensional subspace of X^*. Then there exist $x_1, \ldots, x_m \in X$, with $\|x_j\| = 1$ for all j, such that for all $f \in Y$:*

$$\frac{1}{2}\|f\| \leq \max |f(x_j)|.$$

M.S. Osborne, *Locally Convex Spaces*, Graduate Texts in Mathematics 269, DOI 10.1007/978-3-319-02045-7, © Springer International Publishing Switzerland 2014

Proof. Let $n = \dim Y$, and start with $x_1, \ldots, x_n \in X$ chosen so that all $\|x_j\| = 1$, and $\{x_j + Y_\perp : j = 1, \ldots, n\}$ is a basis of $X/Y_\perp$, *the* dual of Y (weak-* topology). The map

$$f \mapsto \begin{pmatrix} f(x_1) \\ \vdots \\ f(x_n) \end{pmatrix} \in \mathbb{R}^n \text{ or } \mathbb{C}^n$$

is a topological isomorphism (Proposition 2.9), so $\{f \in Y : |f(x_j)| \leq 1 \text{ for } j = 1, \ldots, n\} = \{x_1, \ldots, x_n\}^\circ \cap Y$ is compact. But letting $S = \{x \in X : \|x\| = 1\}$,

$$\{f \in Y : \|f\| \leq 1\} = \left(\bigcup_{\substack{F \subset S \\ F \text{ finite}}} \{x_1, \ldots, x_n\} \bigcup F \right)^\circ \cap Y$$

$$= \bigcap_{\substack{F \subset S \\ F \text{ finite}}} \left((\{x_1, \ldots, x_n\} \bigcup F)^\circ \cap Y \right).$$

If $F \subset S$, with F finite, set

$$K_F = \left(\{x_1, \ldots, x_n\} \bigcup F \right)^\circ \cap Y - \{f \in Y : \|f\| < 2\}.$$

Each K_F is compact, and $K_F \cap K_{F'} = K_{F \cup F'}$. Since $\cap K_F = \emptyset$, there must exist $F = \{x_{n+1}, \ldots, x_m\}$ such that $K_F = \emptyset$. That is, $\{x_1, \ldots, x_m\}^\circ \cap Y \subset \{f \in Y : \|f\| < 2\}$.

Suppose $f \in Y$ and $m = \max |f(x_j)| > 0$. Then $m^{-1} f \in \{x_1, \ldots, x_m\}^\circ$, so $\|m^{-1} f\| < 2$, that is, $\|f\| < 2m$. (If $m = 0$, then $f = 0$ since all $f(x_j) = 0$.) $\square$

Lemma C.2. *Suppose X is a Banach space, and suppose A is a weakly sequentially compact subset of X. Suppose Φ belongs to the weak-* closure of $J_X(A)$, and $f_1, \ldots, f_m \in X^*$. Then there exists $x \in X$ such that $\Phi(f_j) = f_j(x)$ for $j = 1, \ldots, m$.*

Proof. For all n, the set

$$J_X(A) \bigcap \{ \Psi \in X^{**} : |\Psi(f_j) - \Phi(f_j)| < \frac{1}{n} \text{ for } j = 1, \ldots, m \}$$

is nonempty; choose $x_n \in A$ so that

$$|\Phi(f_j) - f_j(x_n)| < \frac{1}{n} \text{ for } j = 1, \ldots, m.$$

The sequence $\langle x_n \rangle$ has a weakly convergent subsequence $\langle x_{n_k} \rangle$ since A is weakly sequentially compact; say $x_{n_k} \to x$. Then $f_j(x_{n_k}) \to f_j(x)$ for all j. But by definition, $f_j(x_n) \to \Phi(f_j)$ for all j, so $\Phi(f_j) = f_j(x)$ for all j. $\square$

Lemma C.3. *Suppose X is a Hausdorff locally convex space and suppose A is a bounded subset of X. Then A is weakly compact if, and only if, $J_X(A)$ is weak-* closed in X^{**}.*

Proof. $J_X : (A$, weak topology$) \to (J_X(A)$, weak-* topology$)$ is a homeomorphism, so if A is weakly compact, then $J_X(A)$ is weak-* compact, hence is weak-* closed in X^{**}. On the other hand, since A is bounded, A° is a strong neighborhood of 0 in X^*, so $A^{\circ\circ}$ is weak-* compact in X^{**} (Banach-Alaoglu). But by definition, $J_X(A) \subset A^{\circ\circ}$, so if $J_X(A)$ is weak-* closed, then $J_X(A)$ is weak-* compact, which makes A weakly compact.   □

Theorem C.4 (Eberlein). *Suppose X is a Banach space, and A is a weakly sequentially compact subset of X. Then A is weakly compact.*

Proof. First of all, A is bounded: Suppose $f \in X^*$. If $f(A)$ is not bounded, then there exists a sequence $\langle x_n \rangle$ in A for which $|f(x_n)| \geq n$. That would mean that $\langle x_n \rangle$ could not have a weakly convergent subsequence (since $|f(x_{n_k})| \to \infty$ for any subsequence), a contradiction. Hence $f(A)$ is bounded for all $f \in X^*$, so A is bounded by Corollary 3.31.

The proof is completed by showing that $J_X(A)$ is weak-* closed in X^{**} and quoting Lemma C.3; Lemmas C.1 and C.2 are used to show that $J_X(A)$ is weak-* closed in X^{**}. If $X = \{0\}$ there is nothing to prove, so assume $X \neq \{0\}$, and Φ belongs to the weak-* closure of $J_X(A)$. Set $Y_1 = \text{span}\{\Phi\}$, and choose $\{f_1, \ldots, f_{n_1}\} \subset X^*$, all $\|f_j\| = 1$, with $\frac{1}{2}\|\Psi\| \leq \max\{|\Psi(f_j)|; j = 1, \ldots, n_1\}$ for $\Psi \in Y_1$. (Lemma C.1. *Note:* We can take $n_1 = 1$, but that doesn't matter.) Choose $x_1 \in A$ so that $f_j(x_1) = \Phi(f_j)$ for $j = 1, \ldots, n_1$, using Lemma C.2. Set $Y_2 = \text{span}\{\Phi, J_X(x_1)\}, \ldots$ In general, given $\Phi, x_1, \ldots, x_m; f_1, \cdots f_{n_m}$, and $Y_{m+1} = \text{span}\{\Phi, J_X(x_1), \ldots, J_X(x_m)\}$, subject to $\Phi(f_j) = f_j(x_m)$ for $j = 1, \ldots, n_m$: Recursively choose $f_{n_m+1}, \ldots, f_{n_{m+1}} \in X^*$, with $\|f_j\| = 1$ all j, for which $\frac{1}{2}\|\Psi\| \leq \max\{|\Psi(f_j)| : j = n_m + 1, \ldots, n_{m+1}\}$ for $\Psi \in Y_{n+1}$; choose $x_{m+1} \in A$ for which $\Phi(f_j) = f_j(x_{m+1})$ for $j = 1, \ldots, n_{m+1}$; and set $Y_{m+2} = \text{span}\{\Phi, J_X(x_1), \ldots, J_X(x_{m+1})\}$.

The preceding produces an ascending sequence of finite-dimensional subspaces Y_m in X^{**}, and a sequence $\langle x_m \rangle$ in A. Since A is sequentially compact, there is a subsequence x_{m_k} which converges weakly to $x \in A$. The claim is that $J_X(x) = \Phi$. There are four parts to this.

First of all, given any j, once $m_k \geq j$ we have that $\Phi(f_j) = f_j(x_{m_k})$, so taking the limit, $\Phi(f_j) = f_j(x)$ for all j.

Second, set $Y = \bigcup Y_m$. If $\Psi \in Y_{m+1}$, then

$$\frac{1}{2}\|\Psi\| \leq \max(|\Psi(f_j)| : j = n_m + 1, \ldots, n_{m+1})$$

$$\leq \sup\{|\Psi(f_j)| : \text{all } j\},$$

so the inequality $\frac{1}{2}\|\Psi\| \leq \sup |\Psi(f_j)|$ holds for all $\Psi \in Y$.

Third, this inequality also holds for $\Psi \in Y^-$, the norm closure of Y: Suppose $\Psi \in Y^-$ and $\varepsilon > 0$. Choose $\Psi_0 \in Y$ for which $\|\Psi - \Psi_0\| < \varepsilon$, and choose j_0 for which $\frac{1}{2}\|\Psi_0\| < |f_{j_0}(\Psi_0)| + \varepsilon$. Then since $\|f_{j_0}\| = 1$:

$$\|\Psi\| = \|\Psi_0 + (\Psi - \Psi_0)\| \leq \|\Psi_0\| + \|\Psi - \Psi_0\| \leq \|\Psi_0\| + \varepsilon \text{ and}$$

$$|f_{j_0}(\Psi_0)| = |f_{j_0}(\Psi) + f_{j_0}(\Psi_0 - \Psi)| \leq |f_{j_0}(\Psi)| + |f_{j_0}(\Psi_0 - \Psi)|$$

$$< |f_{j_0}(\Psi)| + \varepsilon, \text{ giving}$$

$$\|\Psi\| \leq \|\Psi_0\| + \varepsilon < 2|f_{j_0}(\Psi_0)| + 2\varepsilon + \varepsilon$$

$$< 2|f_{j_0}(\Psi)| + 2\varepsilon + 3\varepsilon$$

$$\leq 2 \sup\{|f_j(\Psi)|\} + 5\varepsilon.$$

This holds for all $\varepsilon > 0$, so $\frac{1}{2}\|\Psi\| \leq \sup |\Psi(f_j)|$ for all $\Psi \in Y^-$.

Finally, $J_X(x) \in Y^-$: x is the weak limit of $\langle x_{n_k} \rangle$, so there exists a sequence $\langle y_k \rangle$ for which

$$y_k \in \mathrm{con}\{x_{n_k}, x_{n_{k+1}}, \dots,\}$$

and $y_k \to x$ in norm, by Exercise 20, Chap. 3. But every $J_X(x_n) \in Y$, so $J_X(y_k) \in Y$. Since J_X is an isometry, $J_X(y_k) \to J_X(x)$ in norm.

Now we are done: Y^- is a subspace, and $\Phi, J_X(x) \in Y^-$, so

$$\frac{1}{2}\|\Phi - J_X(x)\| \leq \sup |(\Phi - J_X(x))f_j|$$

$$= \sup |\Phi(f_j) - f_j(x)| = 0.$$

□

The next subject is really the production of a set to which Eberlein's theorem applies. Suppose X is a Hausdorff locally convex space, U is a barrel neighborhood of 0 in X, and K is a weakly compact subset of X. Then U° is a weak-* compact subset of X^*. We are interested in the set $A = \{f|_K : f \in U^\circ\}$ as a subset of $C(K)$, the eventual claim being that A is weakly compact in the Banach space $C(K)$. Unfortunately, the map $f \mapsto f|_K$ from $(X^*, \text{weak-* topology})$ to $(C(K), \text{weak topology})$ need not be continuous, which is why the digression through sequences is necessary. Suppose $\langle f_n \rangle$ is a sequence in U°. The space X, as well as A, K, and $\langle f_n \rangle$, will be as above in Lemmas C.5–C.8.

Lemma C.5. *Every* weak-* *cluster point of* $\langle f_n \rangle$ *belongs to* $U^\circ \cap (\cap \ker(f_j))^\perp$.

Proof. $U^\circ \cap (\cap \ker(f_j))^\perp$ is weak-* closed, and contains every f_j.   □

Now let $\mathbb{F}$ denote the base field, and define

$$\mathbf{f} = \langle f_j|_K \rangle : K \to \prod_{j=1}^{\infty} \mathbb{F}$$

$\prod \mathbb{F}$ is given the product topology, and it is metrizable (Corollary A.4). The function **f** is continuous (Theorem 1.18), so $\mathbf{f}(K)$ is a compact metrizable space. Choose a countable set $D \subset K$ so that $\mathbf{f}(D)$ is dense in $\mathbf{f}(K)$.

Lemma C.6. *There is a subsequence* $\langle f_{n_k} \rangle$ *that converges pointwise on* D.

Proof. Cantor diagonalization: If $D = \{p_1, p_2, \ldots\}$, choose a subsequence $\langle f_{n_{1,k}} \rangle$ for which $\langle f_{n_{1,k}}(p_1) \rangle$ converges; choose a subsequence $\langle f_{n_{2,k}} \rangle$ of $\langle f_{n_{1,k}} \rangle$ for which $\langle f_{n_{2,k}}(p_2) \rangle$ converges, ... Then $\langle f_{n_{k,k}} \rangle$ converges on D.   □

Lemma C.7. *Suppose* f *and* g *are two weak-* cluster points of* $\langle f_{n_k} \rangle$ *in* U°, *where* $\langle f_{n_k} \rangle$ *is as in Lemma C.6. Then* $f|_K = g|_K$.

Proof. By Lemma C.5, both f and g vanish on $\cap \ker(f_j)$, so if $f(x) \neq f(y)$ then $x - y \notin \cap \ker(f_j)$, so some $f_j(x) \neq f_j(y)$. It follows that there exists $F : \mathbf{f}(K) \to \mathbb{F}$ for which $f = F \circ \mathbf{f}$; similarly, $g = G \circ \mathbf{f}$ for some $G : \mathbf{f}(K) \to \mathbb{F}$. These functions F and G are continuous (Corollary A.10), and $f_{n_k}(p) \to r_p$ for $p \in D$ implies that $f(p) = r_p$ since f is a cluster point of $\langle f_{n_k} \rangle$. (Otherwise, if $|f(p) - r_p| = \varepsilon > 0$, then f_{n_k} does not eventually enter $\{h \in X^* : |h(p) - f(p)| < \varepsilon/2\}$.) In particular, $F(\mathbf{f}(p)) = r_p$, and similarly $G(\mathbf{f}(p)) = r_p$. In particular, F agrees with G on $\mathbf{f}(D)$, which is dense in $\mathbf{f}(K)$, so $F = G$. Hence

$$f|_K = F \circ \mathbf{f} = G \circ \mathbf{f} = g|_K.$$

□

Lemma C.8. *If* $\langle f_{n_k} \rangle$ *is as in Lemma C.6, and* f *is a weak-* cluster point of* $\langle f_{n_k} \rangle$ *in* U°, *then* $f_{n_k} \to f$ *pointwise on* K.

Proof. Suppose $\langle f_{n_{k_j}} \rangle$ is a subsequence of $\langle f_{n_k} \rangle$. Then $\langle f_{n_{k_j}} \rangle$ has a cluster point g in U°, and this g is also a cluster point of $\langle f_{n_k} \rangle$, so $f|_K = g|_K$. In particular, $f|_K$ is a cluster point of $\langle f_{n_{k_j}}|_K \rangle$, so $f_{n_k}|_K \to f|_K$ pointwise (Proposition A.6), since pointwise convergence is convergence in a topology (the product topology on $\prod_K \mathbb{F}$).   □

Proposition C.9. *Suppose* X *is a Hausdorff locally convex space,* U *is a barrel neighborhood of* 0 *in* X, *and* K *is a weakly compact subset of* X. *Set* $A = \{f|_K : f \in U^\circ\}$. *Then* A *is weakly compact in the Banach space* $C(K)$.

Proof. In view of Eberlein's theorem, it suffices to show that A is sequentially compact. Suppose $\langle f_n|_K \rangle$ is a sequence in A. Then there is a subsequence $\langle f_{n_k}|_K \rangle$ that converges pointwise on K. Taking $f_{n_k} \in U^\circ$ which give the correct restrictions to K, the sequence $\langle f_{n_k} \rangle$ *does* have a weak-* cluster point f in U° (Proposition 1.5), so that $f_{n_k} \to f$ pointwise on K. Also, K is weakly bounded, so it is bounded (Corollary 3.31), so $K \subset cU$ for some $c > 0$. Hence for $x \in K$, $c^{-1}x \in U \Rightarrow$ all $|f_{n_k}(c^{-1}x)| \leq 1 \Rightarrow$ all $|f_{n_k}(x)| \leq c$.

Suppose $\varphi \in C(K)^*$; then φ is represented by a (signed or complex) measure μ, and we can write

$$\varphi(h) = \int_K h(x)\chi(x)d|\mu|(x), h \in C(K)$$

for some Borel function χ for which $|\chi(x)| = 1$. Since the weak topology on $C(K)$ is "really" a product topology over $C(K)^*$ [the weak-* topology on $C(K)^{**}$ (Proposition 3.24) transported back to $C(K)$], it suffices to show that $\varphi(f_{n_k}|_K) \to \varphi(f|_K)$ for all such φ in $C(K)^*$ (Proposition A.2). But

$$\lim_{k \to \infty} \int_K f_{n_k}(x)\chi(x)d|\mu|(x) = \int_K f(x)\chi(x)d|\mu|(x)$$

by the Lebesgue Dominated convergence theorem (dominated by c). $\square$

We can now prove:

Theorem C.10 (Krein–Smulian II). *Suppose X is a quasi-complete Hausdorff locally convex space. Then the closed convex hull of a weakly compact subset of X is weakly compact.*

Proof. Suppose K is a weakly compact subset of X. The trick involves a judicious choice of topologies. First, define $T : X^* \to C(K)$ by $T(f) = f|_K$. Then by definition, the inverse image under T of the closed unit ball in $C(K)$ is precisely K°, a strong neighborhood of 0 in X^*, so T has an adjoint $T^* : C(K)^* \to X^{**}$. This T^* can be interpreted using integrals, but this is best done after the fact.

Let $C(K)^*_{w*}$ denote $C(K)^*$ equipped with the weak-* topology, and let X^{**}_{w*} denote X^{**} equipped with the weak-* topology.

Letting X_w denote X equipped with the weak topology, note that $J_X : X_w \to J_X(X) \subset X^{**}_{w*}$ is a homeomorphism. Also, $T^* : C(K)^*_{w*} \to X^{**}_{w*}$ is continuous (Theorem 5.2).

The other topology to be considered on $C(K)^*$ is the Mackey topology associated with the weak-* topology, the base being all A°, where A is weakly compact and convex in $C(K)$. Continuous linear functionals in this topology are evaluations at points of $C(K)$. (See Proposition 3.27 and the surrounding discussion.) Let $C(K)^*_M$ denote $C(K)^*$ equipped with this topology.

The other topology to be considered on X^{**} is a new one, the "transported" topology, where a base at 0 consists of all $U^{\circ\circ}$, where U is a barrel neighborhood of 0 in X. If A is bounded in X, then U absorbs A, so A° absorbs U° [Theorem 3.20(f)], that is U° is strongly bounded in X^*, so $U^{\circ\circ}$ is a strong neighborhood of 0 in X^{**}. Furthermore, $U \cap V \subset U \Rightarrow (U \cap V)^\circ \supset U^\circ \Rightarrow (U \cap V)^{\circ\circ} \subset U^{\circ\circ}$. Hence $(U \cap V)^{\circ\circ} \subset U^{\circ\circ} \cap V^{\circ\circ}$, so the conditions in Theorem 3.2 are verified. The topology given by all such $U^{\circ\circ}$ is coarser than the strong topology on X^{**}.

Suppose F is a finite subset of X^*. Then the weak and weak-* topology on $\text{span}(F)$ agree (Proposition 2.9), and $\text{span}(F)$ is a closed (both weak and weak-* by Corollary 2.10), convex, balanced, nonempty subset of X^*, so $(F^\circ)_\circ = (F_\circ)^\circ \subset \text{span}(F)$ by Theorem 3.20(b). But $F_\circ$ is a barrel neighborhood of 0 in X, so

$$(F_\circ)^{\circ\circ} = ((F_\circ)^\circ)^\circ = ((F^\circ)_\circ)^\circ = F^\circ \text{ (bipolar theorem)}$$

is a transported neighborhood of 0 in X^{**}. In particular, the transported topology is finer than the weak-* topology, so it is Hausdorff. Let X^{**}_t denote X^{**} equipped with the transported topology. Then for U a barrel neighborhood of 0 in X, $J_X^{-1}(U^{\circ\circ}) = (U^\circ)_\circ = U$ by Theorem 5.10(b) and the bipolar theorem. Also, $J_X(U) = J_X((U^\circ)_\circ) = U^{\circ\circ} \cap J_X(X)$ by the same two results, so $J_X : X \to J_X(X) \subset X_t^{**}$ is a homeomorphism.

Finally, $T^* : C(K)_M^* \to X_t^{**}$ is continuous, since $T^{*-1}(U^{\circ\circ}) = T(U^\circ)^\circ$ is a Mackey neighborhood of 0 in $C(K)^*$ when U is a barrel neighborhood of 0 in X, by Proposition C.9.

Now let $\hat{K}$ denote those members of $C(K)^*$ that are represented by positive measures of total mass 1. This is a weak-* closed, convex subset of the closed unit ball in $C(K)^*$, since it can be defined by "$\varphi(f) \geq 0$ when $f \geq 0$, and $\varphi(1) = 1$." $\hat{K}$ is therefore weak-* compact. The claim is that $T^*(\hat{K}) \subset J_X(X)$, so that $J_X^{-1}(T^*(\hat{K}))$ is a weakly compact, convex subset of X that contains K, forcing the closed convex hull of K to be weakly compact.

First of all, if μ_p is a point measure at $p \in K$ (i.e. $\int f d\mu_p = f(p)$), then $T^*(\mu_p)(f) = \int f d\mu_p = f(p) = J_X(p)(f)$, so $J_X(p) = T^*(\mu_p)$. In particular, $J_X^{-1}(T^*(\mu_p)) = p \in K$, so $K \subset J_X^{-1}(T^*(\hat{K}))$. Also, $\hat{K}$ is the closed convex hull of its set of extreme points $E(\hat{K})$ (Krein–Milman theorem), and $E(\hat{K}) = \{\mu_p : p \in K\}$ by Corollary A.18. Since $T^*(E(\hat{K})) \subset J_X(X)$, we get that $T^*(\text{con}(E(\hat{K}))) \subset J_X(X)$.

Now take the Mackey closure of $\text{con}(E(\hat{K}))$. This is contained in the weak-* closure of $\text{con}(E(\hat{K}))$ simply because the weak-* topology is coarser, but since the Mackey closure of $\text{con}(E(\hat{K}))$ is convex (Proposition 2.13), it is already weak-* closed by Theorem 3.29. Hence each element μ in $\hat{K}$ is the Mackey limit of a net $\langle \mu_\alpha \rangle$ from $\text{con}(E(\hat{K}))$ (Proposition 1.3(a)). Since $T^* : C(K)_M^* \to X_t^{**}$ is continuous, $\lim T^*(\mu_\alpha) = T^*(\mu)$ in X_t^{**}. Since $\langle T^*(\mu_\alpha) \rangle$ is a convergent net, it is a Cauchy net (Proposition 1.29), and since each $\mu_\alpha \in \text{con}(E(\hat{K}))$: each $T^*(\mu_\alpha) \in J_X(X)$. That means that $\langle J_X^{-1} T^*(\mu_\alpha) \rangle$ is a Cauchy net in X in the original topology. Also, each $J_X^{-1}(T^*(\mu_\alpha))$ is a convex combination of elements of K by earlier considerations, so $J_X^{-1}(T^*(\mu_\alpha)) \in (K^\circ)_\circ$, a bounded set [Theorem 3.20(d)]. Since $\langle J_X^{-1}(T^*(\mu_\alpha)) \rangle$ is a bounded Cauchy net and X is quasi-complete, $\langle J_X^{-1}(T^*(\mu_\alpha)) \rangle$ is convergent. If $x = \lim J_X^{-1}(T^*(\mu_\alpha))$, then $J_X(x) = \lim T^*(\mu_\alpha)$ in the transported topology so $J_X(x) = T^*(\mu)$ by uniqueness of limits. Since μ was arbitrary, $T^*(\hat{K}) \subset J_X(X)$. As noted earlier, this completes the proof. $\qquad \square$

Remark. Since everything in $J_X^{-1}(T^*(\hat{K}))$ is a limit of a net from $\text{con}(K)$, $J_X^{-1}(T^*(\hat{K}))$ actually *is* the closed convex hull of K.

Corollary C.11. *Suppose X is a quasi-complete Hausdorff locally convex space, and suppose K is a weakly compact subset of X. Then $(K^\circ)_\circ$ is weakly compact.*

Proof. $(K^\circ)_\circ$ is closed and convex, so $\text{con}(K)^- \subset (K^\circ)_\circ$. Hence $(\text{con}(K)^-)^\circ \supset ((K^\circ)_\circ)^\circ = K^\circ$. But $K \subset \text{con}(K)^-$, so $K^\circ \supset (\text{con}(K)^-)^\circ$. Combining, $K^\circ = (\text{con}(K)^-)^\circ$. Hence $(K^\circ)_\circ = ((\text{con}(K)^-)^\circ)_\circ$, a weakly compact set by Theorem C.10 and Proposition 3.21. $\qquad \square$

Application: Weak Integrals

Weak integrals for Hilbert-space-valued functions have already appeared, in Chap. 4, Exercise 22. Reformulating, start with a measure space $(K, \mathscr{B}, \mu)$, a Hausdorff locally convex space X, and a function $\varphi : K \to X$ that is **weakly integrable** in the sense that $[f \circ \varphi] \in L^1(\mu)$ for all $f \in X^*$. If X is a Hilbert space, then $f \mapsto \int f \circ \varphi(t) d\mu(t)$ is evaluation at a point of X, since:

1. $f \mapsto [f \circ \varphi]$ is a linear map from X^* to $L^1(\mu)$ which has a sequentially closed graph, so it has a closed graph since X^* is first countable.
2. $f \mapsto [f \circ \varphi]$ is continuous since X^* is barreled (closed graph theorem).
3. $f \mapsto \int f \circ \varphi(t) d\mu(t)$ is evaluation at a point of X since X is reflexive.

In particular, weak integrals can be defined in general for reflexive Banach or LB-spaces. To get anything more general requires some restrictions. First, a generality.

Proposition C.12. *Suppose $(K, \mathscr{B}, \mu)$ is a measure space for which $\mu(K) = 1$, X is a Hausdorff locally convex space, and $\varphi : K \to X$ is a function for which $f \circ \varphi$ is μ-integrable for all $f \in X^*$. Assume $\varphi(K)$ is bounded in X. Then*

$$f \mapsto \int_K f(\varphi(t)) d\mu(t) \tag{$*$}$$

is a strongly continuous linear functional on X^* that belongs to the weak-* *closed convex hull of $J_X(\varphi(K))$.*

Proof. If $f \in \varphi(K)^\circ$, then

$$\left| \int_K f(\varphi(t)) d\mu(t) \right| \leq \int_K |f(\varphi(t))| d\mu(t) \leq 1,$$

so the linear functional $(*)$ maps $\varphi(K)^\circ$ into the closed unit "disk" in the base field, and so is strongly continuous.

If Φ does *not* belong to the weak-* closed convex hull of $J_X(\varphi(K))$, then there exists a weak-* continuous linear functional on X^{**} that separates Φ from that convex hull as described in Corollary 3.11 (and Proposition 3.14): there exists $f \in X^*$ such that $Re\Phi(f) > 1$, but $Re(J_X(\varphi(t))(f)) \leq 1$ for all $t \in K$, that is $1 \geq Ref(\varphi(t))$ for all $t \in K$. Hence $1 \geq \int_K Ref(\varphi(t)) d\mu(t)$, so $\Phi(f) \neq \int_K f(\varphi(t)) d\mu(t)$ and $(*)$ does *not* define Φ. $\qquad\square$

Corollary C.13. *Suppose K is a compact Hausdorff space, μ is a finite Radon measure on K, X is a quasi-complete Hausdorff locally convex space over $\mathbb{F} = \mathbb{R}$ or $\mathbb{C}$, and $\varphi : K \to X$ is a function for which $f \circ \varphi$ is μ-integrable for all $f \in X^*$. Assume that $\varphi(K)$ is contained in a weakly compact subset of X. Then there exists $x \in X$ such that for all $f \in X^*$:*

$$f(x) = \int_K f(\varphi(t)) d\mu(t).$$

In particular, this holds if φ is weakly continuous, that is if $f \circ \varphi : K \to \mathbb{F}$ is continuous for all $f \in X^*$.

Remark. This x is called the **weak integral**, or the **Pettis integral**, of φ. It is written simply as $x = \int_K \varphi(t)d\mu(t)$.

Proof. If φ is weakly continuous, then $\varphi(K)$ *is* weakly compact. If $\varphi(K) \subset K_1$ with K_1 being weakly compact, let K_2 be the closed convex hull of $\varphi(K)$ and K_3 the closed convex hull of K_1. Then K_3 is weakly compact by Krein–Smulian II, and $K_2 \subset K_3$, so K_2 is weakly compact.

Let $c = \mu(K)$. If $c = 0$, then $x = 0$ will do. If $c > 0$, replace μ with $\nu = c^{-1}\mu$ and φ with $\psi = c\varphi$: The closed convex hull of $\psi(K)$ is cK_2, which is compact, and since $\int_K f(\varphi(t))d\mu(t) = \int_K f(\psi(t))d\nu(t)$, by Proposition C.12 the integral defining ($*$) there is given by an element of the weak-$*$ closed convex hull of $J_X(\psi(K))$. But $J_X(cK_2)$ is weak-$*$ compact and convex, and contains $J_X(\psi(K))$, so the integral defining ($*$) is given by an element of $J_X(cK_2)$. That is, there exists $x \in cK_2$ for which (for all $f \in X^*$):

$$f(x) = J_X(x)(f) = \int_K f(\varphi(t))d\mu(t).$$

$\square$

Now for the interpretation of the map in Krein–Smulian II as an integral. Discontinuing the abuse of notation, if μ is the measure representing a member f_μ of $\hat{K}$ in the proof of Theorem C.10, then

$$J_X^{-1}(T^*(f_\mu)) = \int_K t\,d\mu(t).$$

In general, if M is a locally compact Hausdorff space, μ is a Radon measure on M, X is a Hausdorff locally convex space, and $\varphi : M \to X$ is weakly integrable (i.e. $f \circ \varphi$ is integrable for all $f \in X^*$), then

$$x = \int_M \varphi(t)d\mu(t) \text{ \textbf{means}}$$

$$\forall f \in X^* : f(x) = \int_M f(\varphi(t))d\mu(t).$$

The integral is again called a *weak integral* or a *Pettis integral*. Corollary C.13 gives a condition that guarantees its existence in the compact case. A closer examination of the compact case leads to what we need in general. To this end, assume K is compact and μ is a Radon measure on K. In order to approximate $\int_K \varphi(t)d\mu(t)$ with integrals over "large" subsets, the correct analog is Lusin's theorem.

Definition C.14. Suppose M is a locally compact Hausdorff space, μ is a finite Radon measure on M, X is a Hausdorff locally convex space, and $\varphi : M \to X$ is a function for which $f \circ \varphi$ is Borel measurable for all $f \in X^*$. Then φ is **nearly weakly continuous** if for all $\varepsilon > 0$ there exists a compact set $K_\varepsilon \subset M$ such that φ is weakly continuous on K_ε and $\mu(M - K_\varepsilon) < \varepsilon$.

Proposition C.15. *Suppose M is a locally compact Hausdorff space, μ is a finite Radon measure on M, X is a Hausdorff locally convex space, and $\varphi : M \to X$ is a nearly weakly continuous function. If p is a continuous seminorm on X, then $p \circ \varphi$ is Borel measurable on every compact subset $K \subset M$ on which φ is weakly continuous, and $p \circ \varphi$ is measurable on M with respect to the completion μ_c of μ.*

Proof. Suppose φ is weakly continuous on K, and $p \circ \varphi(t_0) > r$. On $\mathbb{R}\varphi(t_0)$, set $g(\lambda\varphi(t_0)) = \lambda p(\varphi(t_0))$ (so that $g(\varphi(t_0)) = p(\varphi(t_0))$). Then $g \leq p$ on $\mathbb{R}\varphi(t_0)$, so g extends to a continuous, real-valued linear functional on X for which $g \leq p$. (Hahn–Banach theorem). If the base field is $\mathbb{R}$, set $f = g$. If the base field is $\mathbb{C}$, choose $f \in X^*$ so that $Ref = g$ (Proposition 3.14). In either case, $Ref \leq p$, and $Ref(\varphi(t_0)) > r$. hence $\{t \in K : Ref(\varphi(t)) > r\}$ is a relatively open subset of K, containing t_0, on which $p(\varphi(t)) > r$. What all this shows is that $\{t \in K : p \circ \varphi(t) > r\}$ is relatively open in K, so $p \circ \varphi$ is Borel measurable on K.

In general, $p \circ \varphi$ is Borel measurable on each compact $K_{1/n}$, where $\mu(M - K_{1/n}) < \frac{1}{n}$, so $p \circ \varphi$ is Borel measurable on $\cup K_{1/n}$. Since $\mu(M - \cup K_{1/n}) = 0$, $p \circ \varphi$ is measurable on M with respect to the completion of μ. $\square$

We are almost there. One last thing, stated in the general form eventually needed:

Proposition C.16. *Suppose M is a locally compact Hausdorff space, μ is a Radon measure on M, X is a Hausdorff locally convex space, and $\varphi : M \to X$ is a function for which $f \circ \varphi$ is μ-integrable on M for all $f \in X^*$. Suppose φ has a weak integral x; that is, suppose for all $f \in X^*$:*

$$f(x) = \int_M f(\varphi(t))d\mu(t).$$

Suppose p is a continuous seminorm on X for which $p \circ \varphi$ is μ_c-integrable, where μ_c is the completion of μ. Then

$$p(x) \leq \int_M p(\varphi(t))d\mu_c(t).$$

Proof. Suppose $p(x) = r$. As in the proof of Proposition C.15, choose $f \in X^*$ for which $Ref \leq p$ and $Ref(x) = r$. Then

$$p(x) = Ref(x) = \int_M Ref(\varphi(t))d\mu_c(t) \leq \int_M p(\varphi(t))d\mu_c(t).$$

$\square$

In defining a weak integral for a nearly weakly continuous φ, some kind of approximation would be needed, and something approaching "absolute continuity of integration" would also be needed. Put together, this leads to the following:

Theorem C.17. *Suppose M is a locally compact Hausdorff space, μ is a Radon measure on M with completion μ_c, and X is a quasi-complete Hausdorff locally convex space. Suppose $\varphi : M \to X$ is a function for which the following happens:*

(i) *For each $f \in X^*$: $f \circ \varphi$ is Borel measurable on M; and*
(ii) *For each continuous seminorm p on X, $p \circ \varphi$ is μ_c-integrable on M; and for each $\varepsilon > 0$, there exists a compact set $K \subset M$ such that $\varphi : K \to X$ is weakly continuous (i.e., for each $f \in X^*$: $f \circ \varphi$ is continuous) and*

$$\int_{M-K} p(\varphi(t)) d\mu_c(t) < \varepsilon.$$

Then: φ has a weak integral. That is, there exists $x \in X$ for which

$$f(x) = \int_M f(\varphi(t)) d\mu(t) \text{ for all } f \in X^*.$$

Proof. First note that if $f \in X^*$, then $|f|$ is continuous seminorm on X, so $f \circ \varphi$ is actually integrable on M. To produce x, set

$$D = \{K \subset X : K \text{ is compact and } \varphi|_K \text{ is weakly continuous}\}.$$

D is directed by set inclusion ($K \prec K'$ when $K \subset K'$) since if φ is weakly continuous on K_1 and K_2, then φ is weakly continuous on $K_1 \cup K_2$. (Look at the inverse image of a weakly closed set.) If $K \in D$, set

$$x_K = \int_K \varphi(t) d\mu(t),$$

which exists by Corollary C.13. The proof will be completed by showing that $\langle x_K : K \in D \rangle$ is a bounded Cauchy net, hence converges to some $x \in K$ (since X is quasi-complete), and this x is the weak integral of φ.

First of all, if U is a barrel neighborhood of 0, and p_U is its Minkowski functional, then by Proposition C.16:

$$p_U(x_K) \le \int_K p_U(\varphi(t)) d\mu(t) \le \int_M p_U(\varphi(t)) d\mu_c(t).$$

Hence all $x_K \in cU$ if $c > \int_M p_U(\varphi(t)) d\mu_c(t)$. Letting U vary: $\langle x_K \rangle$ is bounded. Back to our fixed U, choose $K \in D$ so that

$$\int_{M-K} p_U(\varphi(t)) d\mu_c(t) < 1.$$

If $K_1, K_2 \in D$ with $K_1, K_2 \supset K$, set $K_1 \Delta K_2 = (K_1 - K_2) \cup (K_2 - K_1)$ and $K_0 = K_1 \cap K_2$. Then for all $f \in X^*$:

$$f(x_{K_1} - x_{K_2}) = f(x_{K_1} - x_{K_0}) - f(x_{K_2} - x_{K_0})$$

$$= \int_{K_1} f(\varphi(t)) d\mu(t) - \int_{K_0} f(\varphi(t)) d\mu(t)$$

$$-\left(\int_{K_2} f(\varphi(t)) d\mu(t) - \int_{K_0} f(\varphi(t)) d\mu(t) \right)$$

$$= \int_{K_1 \Delta K_2} f\left(\left\{ \begin{array}{l} 1 \text{ if } t \in K_1 - K_2 \\ -1 \text{ if } t \in K_2 - K_1 \end{array} \right\} \varphi(t)) d\mu(t).$$

Hence by Proposition C.16, since $x_{K_1} - x_{K_2}$ is a weak integral over $K_1 \Delta K_2$:

$$p_U(x_{K_1} - x_{K_2}) \leq \int_{K_1 \Delta K_2} p_U(\varphi(t)) d\mu_c(t)$$

$$\leq \int_{M-K} p_U(\varphi(t)) d\mu_c(t) < 1,$$

so that $x_{K_1} - x_{K_2} \in U$. Again, letting U float, this shows that $\langle x_K : K \in D \rangle$ is a Cauchy net. Setting $x = \lim x_K$, it remains to show that for all $f \in X^*$, $f(x) = \int_M f(\varphi(t)) d\mu(t)$. But this is immediate: If $f \in X^*$, then $|f|$ is a continuous seminorm. For all $K \in D$:

$$\int_{M-K} |f(\varphi(t))| d\mu(t) < \varepsilon, \text{ and } K' \supset K, K' \in D$$

$$\Rightarrow | \int_M f(\varphi(t)) d\mu(t) - \int_{K'} f(\varphi(t)) d\mu(t) |$$

$$\leq \int_{M-K'} |f(\varphi(t))| d\mu(t) < \varepsilon.$$

Since such a K can be chosen, this shows directly that

$$\lim_D \int_K f(\varphi(t)) d\mu(t) = \int_M f(\varphi(t)) d\mu(t).$$

Hence for all $f \in X^*$:

$$f(x) = \lim_D f(x_K) = \lim_D \int_K f(\varphi(t)) d\mu(t)$$

$$= \int_M f(\varphi(t)) d\mu(t).$$

$\square$

The conditions in this theorem are not really standardized. Some use "strongly integrable" for this; others mean something entirely different by "strongly integrable." Two references worth consulting are Edwards [13, Sect. 8.14] and Diestel and Uhl [9, Chap. II].

To close, here is what the Lebesgue dominated convergence theorem looks like in this setting. Note that it is much less satisfactory than in the scalar-valued case, since the limit function has to be assumed to be nice.

Theorem C.18 (Lebesgue Dominated Convergence Theorem). *Suppose M is a locally compact Hausdorff space, μ is a Radon measure on M with completion μ_c, and X is a quasi-complete Hausdorff locally convex space. Suppose $\varphi_n, \varphi : M \to X$ are functions satisfying the conditions imposed on φ in Theorem C.17. Then:*

(a) *If for all $f \in X^*$, there exists a μ-integrable $g : M \to \mathbb{R}$ such that $g(t) \geq |f(\varphi_n(t))|$ for all n and t, and $\varphi_n(t) \to \varphi(t)$ weakly for a.e. $t \in X$, then*

$$\int_M \varphi_n(t)d\mu(t) \to \int_M \varphi(t)d\mu(t) \text{ weakly.}$$

(b) *If for each continuous seminorm p on X, there exists a μ_c-integrable $g : M \to \mathbb{R}$ such that $g(t) \geq p(\varphi_n(t))$ for all n and t, and $\varphi_n(t) \to \varphi(t)$ in the original topology for a.e.t, then*

$$\int_M \varphi_n(t)d\mu(t) \to \int_M \varphi(t)d\mu(t) \text{ (original topology).}$$

Proof. (Outline). Part (a) is just LDCT applied to each $f(\varphi_n(t))$. Part (b) follows from LDCT applied to each $p(\varphi(t) - \varphi_n(t))$ and Proposition C.16; note that $p(\varphi(t) - \varphi_n(t)) \leq 2g(t)$ a.e.   □

Appendix D
Further Hints for Selected Exercises

Chapter 1

12. Use Exercise 11 and induction: If $(G^n)^- = G^{[n]}$, then $[G^{[n]}, G^{[n]}] \subset [G^n, G^n]^- \subset (G^{n+1})^-$. But $G^n \subset G^{[n]}$ as well (also by induction).

17. $d(x^{-1}, y^{-1}) = d(xyx^{-1}, xyy^{-1}) = d(y, x)$ since G is commutative.

18. If K/H is closed, look at its pullback in G. If K is closed, then $G/H - K/H = (G - K)/H$ is open.

20. Open subgroups are always closed. Their complements are the union of their other (open) cosets.

Chapter 2

2. Use Proposition 2.7(e).

Chapter 3

7. W is absorbent, so $(W_\circ)^\circ$ is also absorbent. Hence $W_\circ$ is bounded [Corollary 3.31, condition (iv).] But $(W_\circ)^\circ$ is the weak-* closure of W since W is already convex, balanced, and nonempty. Since V *is* weak-* closed ....

11. Use Fatou's lemma to show that A is closed in $L^p(m)$. (Recall, if needed, that if $\langle [f_n] \rangle$ is an L^p-convergent sequence of function classes, then $\langle f_n \rangle$ has an a.e. convergent subsequence.)

21. Using Exercise 20, choose $h_n \in \mathrm{con}\{f_n, f_{n+1}, f_{n+2}, \ldots\}$ for which $h_n \to f$ in norm. Choose a subsequence $h_{n_k} \to f$ a.e. Do you see why h_n still converges a.e. to g?

23. Not as weird as it looks: $\sum_{n=1}^{\infty} f_n$ is $\sum_{n=1}^{N} f_n$ on X_N. Consult Corollary 3.41.

Chapter 4

4. If A is bounded in Y and $x \in X$, then $\{B(x, y) : y \in A\}$ is bounded in $\mathbb{F}$ since $B(x, ?)$ is continuous. Hence $\{B(?, y) : y \in A\}$ is bounded for pointwise convergence and so is equicontinuous (Theorem 4.16). If U is such that $|B(x, y)| \leq 1$ for $x \in U$ and $y \in A$, then $\{B(x, ?) : x \in U\} \subset A^\circ$.

M.S. Osborne, *Locally Convex Spaces*, Graduate Texts in Mathematics 269, DOI 10.1007/978-3-319-02045-7, © Springer International Publishing Switzerland 2014

10. Much like Theorem 4.24(a), using sequences instead of nets.
19. See Proposition 4.39.
26. Trick Alert: If $U \subset A^-$, then either $U \subset A^- - Y$ or $U \cap (A^- \cap Y) = U \cap Y \neq \emptyset$.

Chapter 5

14. Pull back to X the norm on X^{**} which is dual to $\| |?| \|_*$.
26. If $T(U)^-$ is compact, then $ST(U) \subset S(T(U)^-)$.
29. Also, use Corollary 3.41.

Chapter 6

3a) One approach, which only requires sequential completeness for Z and weak-$*$ closedness (rather than compactness) from each K_n: Suppose U is a barrel neighborhood of 0 in X, and set $A = Z \cap U^\circ$. Verify that Z_A is a Banach space, and use Baire category to (eventually) verify that $A = K_N \cap U^\circ$ for N large enough (beyond where Baire produces an interior point; this is where the "2" in "$2K_n \subset K_{n+1}$" plays its role).

Bibliography

1. Atiyah, M., Bott, R.: A Lefschetz fixed point formula for elliptic complexes I. Ann. Math. **86**, 374–407 (1967)
2. Bachman, G., Narici, L.: Functional Analysis. Dover, Mineola, NY (2000)
3. Bear, H.S.: A Primer of Lebesgue Integration, 2nd edn. Academic Press, San Diego (2002)
4. Boas, R.B.: Invitation to Complex Analysis. Random House, New York (1987)
5. Bourbaki, N.: Topological Vector Spaces. Springer, New York (1987)
6. Bruckner, A., Bruckner, J., Thomson, B.: Real Analysis. Upper Saddle River, NJ, Prentice-Hall (1987)
7. Conway, J.: A Course in Functional Analysis. Springer, New York (1990)
8. De Branges, L.: The Stone–Weierstrass theorem. Proc. Am. Math. Soc. **10**, 822–824 (1959)
9. Diestel, J., Uhl, J.: Vector Measures, AMS Surveys #15, Providence: AMS (1977)
10. Dieudonné, J.: History of Functional Analysis. North-Holland, New York (1981)
11. Dikranjan, D., Prodonov, I., Stayanov, L.: Topological Groups. Marcel Dekker, New York (1990)
12. Dunford, N., Schwartz, J.: Linear Operators, Part I, General Theory. Interscience, New York (1958)
13. Edwards, R.E.: Functional Analysis. Holt, Rinehart, & Winston, New York (1965)
14. Escassut, A.: Ultrametric Banach Algebras. World Scientific, River Edge, NJ (2003)
15. Folland, G.: Real Analysis, 2nd edn. Wiley, New York (1999)
16. Grothendieck, A: Topological Vector Spaces. Gordon & Breach, New York (1973)
17. Hewitt, E., Ross, K.A.: Abstract Harmonic Analysis. Springer, Berlin (1963)
18. Horvath, J.: Topological Vector Spaces and Distributions. Addison-Wesley, Reading, MA (1996)
19. Husain, T.: Introduction to Topological Groups, Philadelphia: W.B. Saunders, (1966)
20. Kelley, J.: General Topology. Van Nostrand, Princeton, NJ (1955)
21. Kelley, J., Namioka, I.: Linear Topological Spaces. Springer, New York (1963)
22. Köthe, G.: Topological Vector Spaces I. Springer, New York (1983)
23. Krein, M.G.: Sur quelques questions de la géometrie des ensembles convexes situés dans un espace linéare normé et complet. Doklady Acad. Sci. URSS **14**, 5–8 (1937)
24. Krein, M.G., Smulian, V.: On regularly convex sets in the space conjugate to a Banach space. Ann. Math (2) **41**, 556–583 (1940)
25. Linderholm, C.: Mathematics Made Difficult. Wolfe, London (1971)
26. Munkres, J.: Topology, 2nd edn. Prentice Hall, Upper Saddle River, NJ (2000)
27. Narici, L., Beckenstein, E.: Topological Vector Spaces. Marcel Dekker, New York (1985)
28. Phelps, R.: Lectures on Choquet's Theorem. Springer, New York (2001)
29. Reed, M., Simon, B.: Methods of Mathematical Physics I: Functional Analysis. Academic Press, New York (1972)

M.S. Osborne, *Locally Convex Spaces*, Graduate Texts in Mathematics 269,
DOI 10.1007/978-3-319-02045-7, © Springer International Publishing Switzerland 2014

30. Royden, H.L.: Real Analysis, 3rd edn. Macmillan, New York (1988)
31. Rudin, W.: Real and Complex Analysis, 3rd edn. McGraw-Hill, New York (1987)
32. Rudin, W.: Functional Analysis, 2nd edn. McGraw-Hill, New York (1991)
33. Schaefer, H.H.: Topological Vector Spaces. Springer, New York (1971)
34. Schechter, M.: Principles of Functional Analysis. Academic Press, New York (1971)
35. Swartz, C.: An Introduction to Functional Analysis. Marcel Dekker, New York (1992)
36. Treves, F.: Topological Vector Spaces, Distributions, and Kernels. Academic Press, New York (1967)
37. Urysohn, P.: Sur les classes ($\mathscr{L}$) de M. Frechet. L'Ens. Math **25**, 77–83 (1926)
38. Van Rooij, A.C.M.: Non-Archimedean Functional Analysis. Marcel Dekker, New York (1978)
39. Wilansky, A.: Modern Methods in Topological Vector Spaces. McGraw-Hill, New York (1978)
40. Wong, Y.-C.: Introductory Theory of Topological Vector Spaces. Marcel Dekker, New York (1992)
41. Yosida, K.: Functional Analysis. 6th edn. Springer, New York (1980)

Index